About Island Press

Island Press is the only nonprofit organization in the United States whose principal purpose is the publication of books on environmental issues and natural resource management. We provide solutions-oriented information to professionals, public officials, business and community leaders, and concerned citizens who are shaping responses to environmental problems.

In 1999, Island Press celebrates its fifteenth anniversary as the leading provider of timely and practical books that take a multidisciplinary approach to critical environmental concerns. Our growing list of titles reflects our commitment to bringing the best of an expanding body of literature to the environmental community throughout North America and the world.

Support for Island Press is provided by The Jenifer Altman Foundation, The Bullitt Foundation, The Mary Flagler Cary Charitable Trust, The Nathan Cummings Foundation, The Geraldine R. Dodge Foundation, The Charles Engelhard Foundation, The Ford Foundation, The Vira I. Heinz Endowment, The W. Alton Jones Foundation, The John D. and Catherine T. MacArthur Foundation, The Andrew W. Mellon Foundation, The Charles Stewart Mott Foundation, The Curtis and Edith Munson Foundation, The National Fish and Wildlife Foundation, The National Science Foundation, The New-Land Foundation, The David and Lucile Packard Foundation, The Pew Charitable Trusts, The Surdna Foundation, The Winslow Foundation, and individual donors.

Unmanaged Landscapes

"A reverence for original landscape is one of the humanities."

—Archie Carr

Unmanaged Landscapes
Voices for Untamed Nature

Edited by Bill Willers

ISLAND PRESS
Washington, D.C. · Covelo, California

ISLAND PRESS is a trademark of The Center for Resource Economics.

Copyright information for previously published material reprinted by permission of the copyright holders appears on pages 233–236.

Library of Congress Cataloging-in-Publication Data
Unmanaged landscapes : voices for untamed nature / edited by Bill
   Willers.
     p. cm
   Includes bibliographical references.
   ISBN 1–55963–693–9 (cloth : alk. paper). — ISBN 1–55963–694–7
(paper : alk. paper)
     1. Fragmented landscapes.   2. Ecosystem management.   I. Willers,
W. B., 1938–
QH541.15.F73U56   1999   99–18456
333.73—dc21   CIP

Printed on recycled, acid-free paper

Manufactured in the United States of America
10 9 8 7 6 5 4 3 2 1

To the Superior Wilderness Action Network (SWAN) and
to its mission to realize a restored wilderness across the northern
forest of Minnesota, Wisconsin, and Michigan

Contents

Introduction   1

Part One: Wildness and Biology   5

"The Question of Management"   7
David W. Orr

"The Nature We Have Lost"   10
Donald Worster

"Is it Un-Biocentric to Manage?"   22
William S. Alverson and Donald M. Waller

From
"Ecology and Revolutionary Thought"   26
Murray Bookchin

From
"Man's Place in Nature"   31
Joseph Wood Krutch

"Antelope"   34
Charles Bowden

From
The Tender Carnivore and the Sacred Game   38
Paul Shepard

"The Balance of Nature"   48
Rachel Carson

"Tsitika to Baram: The Myth of Sustainability"   52
David Duffus

"Toward a Science of Letting Things Be"   56
Bill Willers

"Making Humans Part of the Solution:
A Reply to Willers"   58
Daniel T. Blumstein

"Ecocentrism versus Management:
A Reply to Blumstein"   61
Bill Willers

From
"Down From the Pedestal—A New Role for Experts"   62
David Ehrenfeld

From
"Some Thoughts on Ecological Planning"   64
Raymond F. Dasmann

"A Question and Two Answers"   65
Lucy Redfeather

"What Do We Really Want From the Woods?"   67
Robert Kimber

Part Two: Wildness and Human Society   71

"Killing Wilderness"   73
Wayland Drew

"Resources Everywhere"   81
Wolfgang Sachs

"Bureaucracy and Wilderness (The Grand Design)" 83
Howie Wolke

From
A Wilderness Bill of Rights 102
William O. Douglas

"Travels with Seldom" 103
Ellen Meloy

"The Stone Gallery" 112
Bruce Berger

"Economic Nature" 113
Jack Turner

From
Work in Progress 130
David Brower

From
The End of Nature 132
Bill McKibben

"The Forest of Forgetting" 135
Guy Hand

From
Walden 144
Henry David Thoreau

"A Letter for Montana Wilderness" 144
Rick Bass

"Future Madness" 155
John A. Livingston

"The Good News" 159
Gary Lawless

Part Three: Wildness, Philosophy, and Spirituality 161

From
"The Stranger's Ways" 163
J. Donald Hughes

From
Ecological Revolutions: Nature, Gender, and Science in New England 164
Carolyn Merchant

From
Man and Nature 170
George Perkins Marsh

From
"Wild Wool" 173
John Muir

"Wilderness: A Human Need" 173
Sigurd F. Olson

From
"The Autonomy of Wild Nature" 175
Max Oelschlaeger

From
"Eco-Interests" 179
Lawrence E. Johnson

From
"The Wilderness Idea Reaffirmed" 179
Holmes Rolston III

"The Saving Wildness" 184
Thomas J. Lyon

"Anthropocentrism" 188
John Seed

"The Myths We Live By" 192
George Wuerthner

From
Ways of Nature 194
John Burroughs

"Faking Nature" 194
Robert Elliot

"A Platform of the Deep Ecology Movement" 208
Arne Naess

"The Grace of the Wild" 209
Paul Gruchow

The Imagination of the Earth 211
David Spangler

"Our Life as Gaia" 220
Joanna Macy

About the Editor 225

About the Contributors 227

Acknowledgment of Sources 233

Index 237

Introduction

"We all proclaim our love and respect for wild nature, and in the same breath we confess our firm attachment to values that inexorably demand destruction of the last remnant of wilderness."

—Thomas Merton

Merton had it right. Our culture now effectively functions under an assumption, swept in by the Industrial Revolution and fueled by a Cartesian philosophy of domination, that every part and quality of the planet from pole to pole—land, sea, and air; plant, animal, and mineral—exists for our exploitation and management. While praising wild Nature in the arts and conversation, we nevertheless continue to modify creatures and landscapes for science, technology, agriculture, economics, and recreation, seemingly unmindful of the contradiction between our professed allegiance and our behavior.

Unmanaged landscapes are the focus of the struggle to protect and restore wildness—Nature's autonomy—and to allow for its return on grand scale. At this time of transition between millennia we are beginning to see living systems at all levels—including large ecosystems—in terms not only of parts but also of patterns of internal relationships and processes, with myriad networks nesting within one another, and with feedback loops that self-reproduce, self-organize, self-regulate, create novelty, and evolve. When a living system becomes fragmented or manipulated, its internal pattern of relationships is destroyed. When managed for some human-centered purpose, its autonomy is lost. Restoring wilderness conditions on landscapes of all sizes can allow for self-regulation in a state of ancestral wholeness.

So complex are the inner workings of life's "web" that no one can say for certain how our management schemes for large ecosystems will affect its processes and resident organisms; our best-intentioned operations are launched with insufficient knowledge of the patient. A 1990 report on

1

forestry research by the National Research Council, for example, stated bluntly, "The level of knowledge about forests is inadequate to develop sound forest-management practices." One could say the same about deserts, grasslands, aquatic environments, and alpine ecosystems.

The effects of past and present human activity—logged, mined, and overgrazed landscapes, eroded ozone layer, acid rain, and the like—have by now spread to all parts of the globe, and some will argue that this fact alone is sufficient to counter any arguments for restored wilderness. But those effects do not permanently stifle the natural autonomy—the self-governing capabilities—of unmanaged landscapes, for in the final analysis the healing process itself is a result of internal operations proceeding independently. If land is *maintained* for human-centered purposes, its internal mechanisms are no longer in control and, therefore, are not what are determining evolutionary direction. On the other hand, neither human-induced phenomena, nor nonhuman impacts such as geologic or climatic change, take permanent control of a landscape's internal ability to heal and to adjust to new conditions.

Exactly what constitutes "management" continues to be a subject of debate among wilderness advocates. Where wilderness is to be restored, some insist on a sudden and complete hands-off policy. But this ignores the fact that many landscapes have become so fragmented, and their floral and faunal assemblages so diminished or displaced by aggressive exotic species, that a return to anything resembling a normal ancestral wholeness is no longer possible without some intervention. But intervention geared toward the return to a biologically diverse ancestral condition of self-governance—the essence of wildness—is the opposite, both philosophically and in practice, of the resource-oriented management one sees, for example, in the culturing of favored game animals, such as deer, or of commercially-valuable trees, such as aspen. While the latter yields and maintains something unlike historically normal conditions, restoration is geared toward a return of autonomy rather than toward forcing a set of conditions. The ideal of restoration is to make itself, in the long run, unnecessary.

Over the years, there has been no lack of dissenters from the managerial assumption that has dominated Western culture for so long. The forty-four entries that make up this anthology, which were written from the mid-nineteenth century to the present and which range in length from a few paragraphs to full-blown essays, all argue, directly or indirectly, for Nature's autonomy. Although any given author may contend from several perspectives, there is generally enough focus of argument with each essay to allow the collection to be divided into three parts. Part One examines wildness from the standpoint of biology and ecology, Part Two looks at Nature's autonomy as it relates to human society and its economic concerns, and Part Three considers the philosophic and spiritual aspects of wildness.

Pastoral and managed landscapes have their place in a world where people reside, but like cities they are akin to domesticated animals, molded into entities to do our bidding. If wildness and wild creatures are to survive on Earth, so then must unmanaged landscapes, for they are the fountainheads of the wildness that Henry Thoreau taught is the preservation of the world. They are the blank spots on the map longed for by Aldo Leopold.

Before planetary healing can begin, we will have to reawaken to the priceless quality that is wildness at grand scale. As Sir Peter Scott wrote, "We should have the wisdom to know when to leave a place alone." Reawakening to wildness will carry with it deference to the dense meshwork of process that has been the source of evolution since life on Earth began, and an acknowledgment of the inherent rightness of allowing some vast landscapes to function, with all parts intact, according to their own internal dictates.

Part One

Wildness and Biology

"Wilderness holds answers to questions man has not yet learned how to ask."

—Mary Newhall

Despite legislative mandates, rising public concern, and bureaucratic planning from the federal level down to the county, the fragmentation of landscapes is ongoing, as are the coinciding loss of biological diversity and wild characteristics. Some "resource science" has fostered the alteration of vast and complex systems by managing specifically for sundry industrial and recreational interests—aspen monocultures for paper, plantation pines for timber, disproportionally large populations of deer for hunters, grasses for the domestic livestock that displace native creatures, immense networks of roads for the masses—and this with insufficient attention to the fundamental scientific principle of experimental control: the establishment of parallel situations as standards of comparison. From a scientific standpoint alone, the need for baselines of self-governing systems is reason enough for the protection of unmanaged landscapes.

"The Question of Management"
David W. Orr

There is a growing presumption that humans must now take control of the planet and its life processes. The September 1989 issue of *Scientific American* devoted an entire issue to the subject of "Managing Planet Earth." In it one finds pleas for "adaptive" policies for planetary management that require "improving the flow of information," "technologies for sustainable development," and "mechanisms . . . to coordinate managerial activities" (Clark 1989). Economists of similar mindset intend to "find the right policy levers" to manage "all assets, natural resources, and human resources" (Repetto 1986). Language like this conjures images of economists and policy experts sitting in a computerized planetary control room, coolly pushing buttons and pulling levers, guiding the planet to something called "sustainable growth."

I do not doubt that something needs to be managed. I would like to raise questions, however, about what and how we manage. For would-be planet managers, it is a matter of no small consequence that God, Gaia, or evolution was doing the job nicely until human population, technology, and economics got out of control. This leads me to think that it is humans that need managing, not the planet. This is more than semantic hairsplitting. "Planetary management" has a nice ring to it. It places the blame on the planet, not on human stupidity, arrogance, and ecological malfeasance, which do not have a nice ring. The term avoids the messy subjects of politics, justice, and the discipline of moral choice. Planetary management, moreover, appeals to our desire to be in control of things. It appeals to our fascination with digital readouts, computer printouts, dials, gauges, and high tech of all sorts. Management is mechanical, not organic, and we like mechanical things, which reinforce our belief that we are in control.

Plans to manage the Earth are founded on the belief that ignorance is a solvable problem, that with enough research, satellite data, and computer models, we can take command of Spaceship Earth. There is a great deal more to learn about the Earth, no doubt. But there are good reasons to believe that its complexity is permanently beyond our comprehension. A square meter of topsoil several inches deep teems with life forms that

Repetto, R. 1986. *World enough and time.* Yale University Press, New Haven, Connecticut, p. 8.

✳

"The Nature We Have Lost"
Donald Worster

Nostalgia runs all through this society—fortunately, for it may be our only hope of salvation. My own version, which I probably share with a few million others, takes me back to walk in pristine natural places on this continent. I dream of traveling with our second native-born naturalist, William Bartram (his father John was the first), a slightly daft Pennsylvania Quaker who botanized from the Carolinas down into Florida in the early 1770s. I would travel with him, "seduced by . . . sublime enchanting scenes of primitive nature," through aromatic groves of magnolia, sweet gum, cabbage palmetto, loblolly pine, live oak, the roaring of alligators in our ears. I would gaze with Thomas Jefferson through his elegant white-framed windows at Monticello toward the Blue Ridge Mountains, speculating about the prodigious country stretching west. Best of all, I imagine entering that west with Lewis and Clark in 1804–5, standing beside them on Spirit Mound in present-day South Dakota, beholding, as Clark put it in his execrable spelling, "a most butifull landscape; Numerous herds of buffalow were Seen feeding in various directions; the Plain to North N.W. & N.E. extends without interuption as for as Can be seen." And I think what it must have been like for them warping and poling up the muddy Missouri River, penetrating farther into the vast open country of the unplowed, unfenced prairies when wolves still howled in the night; of heading into "the great unknown," panting over the unpainted, unmined, unskied Rocky Mountains and rafting down the uncharted, undammed Columbia to the gray-green drizzly shore of the Pacific Ocean.

How much has been lost in our short years as a nation, how much have we to be nostalgic about. In the beginning of white discovery North America must have been a glorious place, brimming with exquisite wild beauty, offering to agriculturalists some of the earth's richest soils, incredible stands of trees, booty on booty of mineral wealth. Think for a moment of the infinitude of animals that once teemed but are now diminished or gone.

In the most comprehensive, detailed analysis yet offered, Frank Gilbert Roe estimated that forty million bison roamed the continent as late as 1830. One of the first Europeans to see them, the Spanish explorer Francisco Vasquez de Coronado, wrote almost three hundred years before that date: "I found so many cattle . . . that it would be impossible to estimate their number. For in traveling over the plains, there was not a single day, until my return, that I lost sight of them." So impressed by this animal were later Americans that they put its picture on one of their most common coins; now there are far more of those nickel images saved by coin collectors than bison that survive.

Ernest Thompson Seton estimated forty million white-tailed deer before there were farms and guns. Someone else has said there may have been five billion prairie dogs, as many as the present total human population of the world. And as many as three to five billion passenger pigeons, migrating in dark, torn clouds that blotted out the sun, breaking trees when they came down to roost; now they too have vanished into carbon and gas.

Navigators encountered off Newfoundland schools of fish so dense they blocked their passage, holding them prisoner, and waterfowl so thick they could feast forever on wild duck eggs. In 1985, however, as one index of change, the U.S. Fish and Wildlife Service counted only 62 million among all the major duck species, down more than half from a few decades earlier. If that seems like plenty of ducks, remember that we have about as many tennis rackets in our closets and far more beer cans in our refrigerators. In this year there are over a thousand species on the endangered list, and many more are threatened.

What we have left in abundance, rivaling ourselves in magnitude, are the bugs. We have not really dented their numbers; 10,000 of them per acre can still be found in some American habitats, along with billions of mites, whole Manhattans, whole miniature Chinas, crawling underfoot.

Besides losing so many of the larger animals, we have lost entire ecological communities, complete landscapes, and with them have lost a considerable range of human feelings—the delight and the joy, the humility that may come from standing in the presence of what we have called the wilderness. In most parts of the country such feelings are gone forever.

Environmental historians debate whether the wilderness ever really existed or whether it was merely a figment of the astonished, naive European imagination. Several have argued that the white man invaded a long inhabited, and much managed, country, the home of Indians for tens of thousands of years. North America, they say, was not a "virgin" land but a "widowed" one, as millions of aborigines died from contact with European diseases. But neither adjective will quite do, for the continent was far too big and diverse to be so simply gendered and personalized. It was neither a widow nor a virgin, neither bride nor groom, nothing like any person we have known. But we may accurately say it was, over most of its extent, an untrammeled land by the standards of either early modern western Europe or today's America. Immense stretches of it seldom knew a human foot, let alone a fence or building. The most plausible current estimate of the native human population living north of Mexico in the era of Christopher Columbus runs about five million—extreme figures go as high as eighteen million. Either figure is tiny, compared with all those bison or prairie dogs or compared with most nations today. They had all to themselves a territory of 7.4 million square miles, the combined extent of the United States and Canada, or more than a square mile apiece. In some places, of course, like Florida, the coastal fringe of the Pacific Northwest, the lower Mississippi valley, they lived in much denser populations than that—but then over some part of the landscape virtually no one must have lived: zero humans per square mile. And however numerous or scarce they were or however distributed, they were a Stone Age people, living by hunting and gathering or, where they were agricultural, cultivating their scattered, shifting fields with bones and digging sticks; by far their most potent technology was fire, which they used liberally but undoubtedly controlled even less effectively than we control our nuclear reactors and pesticides. To describe their relationship with the whole continent as "management" would be a considerable exaggeration. Without bogging down in pedantic wrangles over definitions, we can say that before contact the native peoples were dwelling on a largely undomesticated continent, wild or nearly wild over much of its extent. In contrast, the newcomers, the white and black and Asian immigrants, have, in the space of two or three centuries, steadily cleared and paved, mowed and malled their way through that territory, until today less than 2 percent of the lower 48 states is in a condition of wilderness, roadless and uninhabited.

What was it that destroyed so much of America's wild beauty and the Indian's habitat? What turned so much of the continent into the agricultural-urban-industrial-commercial excrescence we see sprawling everywhere today—into Greater Cleveland; the Las Vegas Strip; Harlan County, Kentucky; the Great Plains wheat belt? Merely saying it was the coming of "new immigrants" does not explain very much. What was it about those immigrants that produced such Promethean effects? This is a question that environmental historians have sought to answer, though they have not arrived at any consensus. One set of explanations they offer emphasizes tangible, material forces, including such things as demography, technology, and energy. These, many say, have been the revolutionary forces, and we, like the land on which we live, have been inextricably held in their grip. Let us give those forces their due and briefly recall how they have operated to transform the ecological conditions of the continent.

First, consider the explosive growth in North America's human population following in the wake of the European discoveries. Certainly, and it is obvious to everyone, this has been a massive reason for the dramatic environmental changes that have occurred. The recording of immigration to the United States began only in 1819, two centuries after Jamestown was founded and eleven years after the end of the African slave trade. From that beginning of official documentation until 1970, over 45 million persons legally entered our doorways from abroad. The high annual point was 1907, when 1,285,349 of them were registered. Since 1970 the gate has been opened wide again, and Congress has just recently, in the immigration law of 1990, opened it even wider, allowing in 770,000 per year. Many more, of course, crawl under the fence and enter illegally from Mexico or Canada. Only about 30 percent of the nation's demographic growth now comes from legal immigration, most of the rest from natural increase, but immigration alone adds every decade more than the aboriginal Indian numbers I mentioned above. And now in the late twentieth century we have dwelling in this country over 250 million citizens—an astounding quarter of a billion. That gives us an average density of about seventy persons per square mile, compared with a global average (leaving out the land mass of Antarctica) of one hundred per square mile—and remember that there may have been less than one person per average square mile in Indian North America. The places where most Americans live are now among the most densely settled on Earth. The state of Virginia, for example, has 144 residents per square mile, compared with Iraq's 83. The state of Pennsylvania is more densely peopled than France, with 264 per square mile. Ohio is more thickly settled than Thailand or China. Oklahoma has a greater human density than Finland, Sweden, or Norway. Florida exceeds the density of Spain, Massachusetts exceeds that of India and the United Kingdom, and New Jersey has a hundred more people per square mile than the Netherlands.

Those numbers, massive though they are, account for only a small part of American's total ecological impact. Historians go on to stress a second material factor, the growth of technology and industrialism and what they have done to the biophysical environment. The rise of mass production processes and of mass consumption, starting with the manufacture of cotton textiles in Massachusetts during the 1790s, has completely revolutionized people's relation to the land. Industrialism has filled our wardrobes, larders, and bookcases, filled our households, offices, and barns, with a quantity of goods that come to us almost effortlessly, and come from every part of the country and beyond, from places we have never seen or have little interest in. Since 1914, when Henry Ford set up his assembly lines at the new, advanced River Rouge plant near Detroit, it has been the automobile that has most dramatically symbolized the industrial economy; now, some 100,000 automobiles roll off the world's assembly lines each day, and there are well over a hundred million passenger cars on our nation's streets and highways, as well as millions of trucks and vans. Throughout America and the rest of the industrial world, the automobile has become the most common and the most potent technological force for environmental modification and destruction we have, voraciously consuming wood, aluminum, steel, rubber, plastic, farmland, city space, wetlands, quiet.

And the proliferation of the automobile suggests a third material factor, the rise of the fossil fuels, which have likewise profoundly altered the ecological conditions of the continent. By discovering and tapping those ancient buried stores of energy, coal, oil, and natural gas, we have unleashed incredible changes on the face of the earth here and abroad. In the mid-nineteenth century the average northern household required for fuel sixteen to eighteen cords of wood a year. Today, Americans produce commercially over 58,000 petajoules of energy per year—the equivalent of 9.5 billion barrels of oil—or 38 barrels apiece. A family of four would thus use 152 barrels annually or their equivalent—those are barrels, not gallons. Imagine that today we got our energy supply delivered once a year and had to store it in the backyard, as we once did our fuelwood. Think of all the tidy suburbs of the nation with black barrels of oil plunked down on the neatly mown bluegrass lawns. Never mind for the moment that much of our energy supply is lost in the form of heat, doing us no good whatsoever, or that so much of it is wasted that often the real purpose of fossil fuel consumption seems to be to vent excess CO_2 into the atmosphere. Despite the waste, the continent has been radically transformed during the fossil fuel era. Put those barrels to work on remaking the landscape, as they are every day, every year, and there is no mountain or river that can defy them, if the human will is there, no species that can run away. If you doubt that power try building an interstate highway with an elk's shoulder blade tied to a stick.

So we have to acknowledge those powerful material forces and their history in order to explain the radical transformation of North America since the days of Bartram, Jefferson, Lewis and Clark. But most environmental historians urge caution and even skepticism about letting such materialist explanations become the whole story. Demography, technology, and energy seem impressive forces, all right, but they are also inadequate as explanations, often concealing as they do the real forces of change that lie more hidden and subtle than a coal seam. Historians say it is in the area of the immigrant white man's culture, his and her ideas or ways of seeing, that we can locate the profoundest causes of environmental change. Even the runaway force of American population growth, if scrutinized carefully, is a reflection of cultural attitudes. We have those mounting numbers to deal with not simply because of reproductive biology but because of our common perceptions of what the country's environment will sustain, what it's good for. Why does the United States continue to admit more immigrants than all the other nations of the world? Why does New Jersey vote to admit more people even though it's more crowded than Japan? Because the great American fertility goddess, the Statue of Liberty, keeps whispering in our ears that there's always room for more, that somewhere out in North Dakota there's plenty of space left, that this is and must be the opportunity society. It's a cultural thing.

So let us return to some deeper cultural sources to explain the destruction of nature that has gone grinding along, so inexorably, so fatally, in North America. The most adequate explanation for that destruction (or should we call it "settling the place," as we once thought we were doing?) lies in American attitudes toward nature and our place in it. Those attitudes, I believe, took indelible form in the eighteenth century, the very period when the nation itself emerged and began seeking an identity, forming national ambitions, and planning for the future. We keep harking back to that period because it was then that we ceased to be Europeans and set in motion those distinctive, identifying ideas and institutions that would bring us down to the present. To be sure, those ideas had roots going far back into an archaic past, all the way back to the Greeks and Romans and the ancient Israelites. But it was not until the eighteenth century that the ideas came together into a recognizably modern form, becoming an intellectual mold that has fashioned our national thinking ever since. Those ideas would have cataclysmic ecological consequences; they would drive us relentlessly to create the man-made landscape we inhabit today and in the process nearly wipe out that wilder America.

They would also leave us, for many intricate reasons, feeling guilty about what we have done and would encourage that peculiar search for national atonement we have called environmentalism. All in all a fertile, potent set of ideas. And they came boiling out of quaint little villages like Colonial Williamsburg and Puritan Boston, from taverns redolent with the musty odor of ale, from solemn assembly rooms where politicians orated, from domestic hearthsides where they were spun out with the thread on a spinning wheel.

The key American environmental idea, and at once the most destructive and most creative, the most complacent and most radical, is the one that ironically has about it an aura of wonderful innocence. America, we have believed, is literally the Garden of Eden restored. It is the paradise once lost but now happily regained. In Judeo-Christian mythology the first humans, Adam and Eve, discovering evil after yielding to the Devil's temptation, had to be kicked out of the Garden on their nearly naked bums. But *mirabile dictu,* Americans of the eighteenth century found a way to sneak back into the garden. A band of their ancestors had made their way to the New World and there rediscovered it, with the gate standing wide open, undefended. What a blessed people. They brought along with them some Africans in chains to help enjoy the place, and by and by they let in a few others from Asia, but mainly it was a fortunate band of white Europeans that destiny allowed to reenter and repossess the long-lost paradise. No other people in the world has ever believed, as Americans have, that they are actually living in Eden: not the Italians or the Swiss, though they probably thank their stars for who they are; not the Brazilians or the Mexicans, though they have been living like us among the riches of the New World; not even the Australians or the Canadians, who like us have conquered a lot of valuable real estate; and of course not any of those people who have lived outside the Judeo-Christian tradition, the Koreans or Tibetans, for example. All of them undoubtedly have loved their land and loved it deeply; they have praised its beauty, identified its sacred places, but they have no grand, mythic illusions about it. Many times they have even felt compelled to apologize to the world for the poverty of their soil or the paucity of their resources. Not so with the Americans; we understood early on that the planet's last best place had been kept sequestered for us. That mythic belief in Eden restored lies at the very core of our peculiar national identity. It is the primary source of our self-confidence and our legendary, indefatigable optimism. It has made us the Teflon-coated people, completely impervious to the stark, damning evidence of our own folly. And it has encouraged a rather condescending, pitying attitude toward the rest of the world, who are so sadly condemned to live in such manifestly inferior settings.

There are a couple of themes densely embedded in America's Edenic image of itself, and I want to pull them apart for analysis. The first is the belief that nature, as found in North America, is a complete, eternal, and morally perfect order. To be sure, nature displays the same general hand-icraft in every corner of the planet, but only here is the perfection of her workmanship supposed to be so clearly realized. At some ancient point in time, eighteenth-century Americans believed, the natural world was cre-ated by a supreme intelligence, someone I can't help picturing as a kind of divine Albert Einstein figure, incredibly smart and wonderfully kind, whose frizzy halo of hair has been pressed down under a powdered wig. When he got done creating, he shuffled off and sat in the corner, leaving nature to roll marvelously round and round forever, finished and invul-nerable and beautiful beyond flaw. Obviously, much of that notion can be traced back to Moses (or to whomever she was who wrote the book of Genesis) but the salient point for Americans was that we are privileged to witness that divine creation as perfectly intact as it was at the beginning of time, and what's more, we can actually walk through it.

So tell us, fortunate American citizen, exactly what the properties of your Eden are. Without hesitation, the answers come surely and easily, the familiar stuff of many eighteenth-century writings. We may imagine, for instance, the authoritative voice of Thomas Jefferson answering the ques-tion. To begin with, he would reply, nature is eminently rational, laid out by Great Reason itself and accessible to our lower faculties. It is completely transparent to scientific, inquiring minds, though its creation required more rationality than a whole college of professors could furnish. Furthermore, nature is permanent; it is designed to endure forever as the exact same order of objects interacting with one another in the same old way, surmounting all the vicissitudes of time. Individual plants and ani-mals may come and go, but the order of the whole cannot be altered by any force, including humans acting with all their power. "Such is the econ-omy of nature," wrote Jefferson in *Notes on the State of Virginia*, "that no instance can be produced, of her having permitted any one race of ani-mals to become extinct; of her having formed any link in her great work so weak as to be broken." He was writing to refute rumors that a great-clawed sloth had once walked the North American continent but was now gone. The Indians, Jefferson pointed out, told of such an animal still prowling the northern latitudes, far from human eyes, but their testimony was superfluous; for him abstract reason alone was enough to establish the creature's survival. How could America be the perfect garden if sloths went about dying in it? Show me a supposedly fossil bone and I'll find the living animal to match it, even if I have to paddle my canoe to darkest Arkansas. Another characteristic of nature is that she is frugal and efficient; nothing is ever wasted or useless in her workings. Then, and perhaps most important, nature is amazingly abundant. Every tiny niche is filled with life, and each little life contributes to the welfare of the whole. Here is Thomas Paine's enthralled account of that cornucopia:

> If we take a survey of our own world . . . our portion in the immense system of creation, we find every part of it, the earth, the waters, and the air that surround it, filled, and as it were crowded with life, down from the largest animals that we know of to the smallest insects the naked eye can behold, and from thence to others still smaller, and totally invisible without the assistance of the microscope. Every tree, every plant, every leaf serves not only as an habitation, but as a world to some numerous race, till animal existence becomes so exceedingly refined, that the effluvia of a blade of grass would be food for thousands.

In the face of such plentitude, the human being can only consume in a spirit of gratitude whatever he or she can. Nature demands our respect, but how do we show it? By the degree of our enthusiasm, the intensity of our enjoyment, the size of our appetite. We are not required to pinch and save, to conserve or protect, because some one much more intelligent has done all that for us. Nature will look out for itself. Our only work is to be active in making good and full use of the abundance spread before us. Could anything be more Edenic than that?

The power of the eighteenth-century myth of an American Eden reverberates down through our history to these very times. We are still a people in love with our prolific Garden. We continue to explore it, as Jefferson's generation did, traveling with pride and satisfaction from valley to mountain to plain, fencing in what we can, owning and possessing as much as we can, mining and exploiting it with great glee, now and then settling down to enjoy the view. We demand maximum freedom to drive our new Buick (advertised under the Edenic sign of the American eagle) to and through the garden, bumper to bumper with our Nissan Pathfinders, our Ford Broncos, our Vagabond motorhomes. We have scattered Edenic images all across the country's map, from Garden City, Long

Island, and the Garden State of New Jersey to Garden City, Kansas, and the Garden of the Gods, Colorado. In our national gazeteer you can find an Eden, Michigan, an Eden, Mississippi, an Eden, New York, an Eden, Wyoming, as well as an Eden Valley and Eden Prairie in Minnesota. Many of them may be marred these days by a Kmart, a Wendy's, an overflowing landfill, or by air pollution from a nearby smelter; never mind, the names on the map often sounded more glamorous than the reality looked.

Americans, even in these days of rising concern, still tend to look on nature through an eighteenth-century rose-colored window—Palladian is understood. I speak not only of the creationists, who still consider Moses the best biologist ever and vehemently reject the overwhelming testimony of scientists about evolution, but also of economists, politicians, manufacturers, and happy summer campers, for all of whom nature is forever infinitely generous, forgiving, and abundant. We cannot do any real damage to her, we still say to ourselves; she is now, as ever, a mother who never says no to her children. Eden survives for us, if only in the endlessness of our material expectations.

Ironically, many environmentalists today, though disagreeing with so much of the thinking I have just described, also hold fast to some eighteenth-century notions of nature. They look to the last remaining wilderness for proof of a benevolent moral order existing outside human culture, a standard by which the fall of humankind may be measured. Such is the essential creed, I believe, of groups like the Wilderness Society and the Sierra Club, though for them that natural Eden is no longer protected or secured by divine intelligence, nor is it infinitely abundant, able to satisfy all our demands. Eden has been much spoiled in their eyes, and reversing the tables, it is now up to us to redeem what is left of it.

As these environmentalist revisions indicate, the whole idea of an invulnerable Eden waiting through the ages on the North American continent has been slowly breaking down for a long time now. Even before Thomas Jefferson died in 1826, scientists had firmly established that many species had truly gone extinct, leaving large holes in the natural order. The publication in 1859 of Charles Darwin's *On the Origin of Species* only added to the sneaking awareness that we cannot actually go back to some primeval Eden, if it ever existed in quite the way Jefferson's age believed. Today, most well-informed Americans readily grant that we live in the midst of a nature that has never been exactly perfect, not for the last hundred million or hundred billion years, a nature that has always been evolving and changing, though we may agree that it has shown an impressive trial-and-error adaptativeness, far more than our recently contrived economy or laws. The nature we now see looks far more fragile and vulnerable than it once did, less efficient in many ways, and far less bountiful in light of the demands we make on it. We are moving at long last, even on this uniquely favored continent, toward an awareness of universal resource scarcity and limits.

How much enthusiasm can we work up for this less than perfect nature revealed by the modern era? Is reverence really possible any longer? Is nature in this post-Edenic era still worthy of great respect, or should we put more faith in our own human intelligence, in our technological inventiveness? Must our hopes now rest not on the land but on our ingenuity, and if so, is our ingenuity really up to the challenge? Conversely, can we now ignore the need to accept responsibility for the environmental damage we do? Such ethical questions did not puzzle our ancestors, who were serious moral thinkers but seldom thought they had any moral obligations to the earth. They plunge us back into the dark, tangled, disordered web of history we have made.

Thus the image handed down to us from the eighteenth century of America as a restored Eden is in a most confused state these days—not completely discredited but less credible than before. That image is sharply contradicted by modern experience and more and more seems downright complacent and naive. One thing is clear to the historian: out of our resulting intellectual confusion we are growing toward a more realistic sense of what the natural world will allow us to do.

But there is a second theme in our national myth of Eden that holds on quite tenaciously still, driving us unwittingly to do further havoc to ourselves and the earth: the notion that human nature in America is a mirror image of that perfect garden of nature and is itself benevolent and good. Supposedly, we live in a society that, though admittedly not yet perfect, is moving in that direction. We are humanity's last best hope. We can trust in ourselves if not in the rest of nature.

To believe, as we have long done, that America will provide the world with a model society, is a bold, ambitious, and arrogant faith. It has implied that we are a people who can bring off an unprecedented moral revolution in human character, one that all the previous history of the species failed to do. That revolution, Americans announced as early as the eighteenth century, was already beginning to happen; under the benevolent influence of New World nature, the old vices and evils of humanity were fast disappearing. Take greed, for example. Every major religious teacher of the Old World had denounced greed as one of the cardinal sins. Jesus, for instance, had declared that it was harder for a rich man to enter the Kingdom of Heaven than for a camel to squeeze through the eye of a needle. Apparently in America, however, our camels were smaller and our needles larger. In this fortunate place we maintained that human greed could actually metamorphose into a virtue. The more we freed greed from laws and regulations, the more we would prosper; and the more we prospered, the more we would become enlightened, happy, and decent. This is, of course, the fundamental moral argument we associate with the rise of capitalism and the market economy, the dominant economic philosophy and institution of the modern world. Capitalism, in order to become so dominant, had to convince people, against all the weight of tradition, that greed was truly a virtue. That transvaluation may have owed much of its theoretical argument to an Edinburghian economist named Adam Smith, but it found the greatest receptivity, its most zealous believers, among Americans. In other countries the promoters of capitalism had to fight harder against such traditional opponents as the Christian Church, older social elites, and some intellectuals and artists, who tended to be suspicious and critical. But then that hostile opposition was not living in Eden and found it hard to believe in the actuality of moral revolution; the greedy, they insisted, still needed, as always, to be watched and restrained. In contrast, Americans tended to dismiss all such suspicion of human nature as outmoded and unnecessary. Everybody might become innocent in this providential environment—at least all the well-dressed white guys. And so it is today. Our rich men are as well esteemed as ever and have not only entered into but are in control of our secular heaven, from Wall Street to Malibu Canyon. No matter how they got their money, we still tend to admire and praise them, and to forgive their sins easily. Even the rogue serving ten years in the slammer for bilking widows of their life savings, or for stealing from public housing funds, or for conning his television audience into sending in donations can do no wrong in some eyes. He loudly proclaims his innocence and virtue—"I was only trying to do my job, I am not a crook"—and millions of Americans believe him.

This notion that America is the land where vice is transformed into virtue, where everybody is as innocent as a child, has admittedly begun to wobble a bit, but it is still strong in our thinking, particularly in our economic thinking, which is about all that our thinking amounts to much of the time. Every announcement of higher economic growth is greeted as proof of our improving virtue and idealism, rather than of our insatiable, ruinous greed. Despite plenty of evidence that selfishness is as ugly and rampant here, and the lust for power as corrupt, as in other nations, despite many examples of the fact that Eden has not cleansed or restored us to innocence, we resist the logical conclusion that we need more law, regulation, and good government in the national political economy as in the private domestic one. We resist that conclusion when we elect men to high office whose only qualification is their fervent, unquestioning belief that we are still the unblemished moral light of the world—if only the government would get out of the way.

Will this charming set of self-images ever seem as naive as Jefferson's theory that animal species can never become extinct? Will we ever decide that the hunger for unlimited wealth and power is as dangerous in the New World as in the Old? Will all of us ever come to see and regret the impact that unrepressed human appetite has on the natural environment? Regaining paradise is an ancient, wonderful dream, and the Edenic dream and the Edenic nature that inspired it are the most wonderful things about America. They explain why we are a nation of idealists. They explain the fact that we are an avid group of nature lovers, as our snapshots, or coffee table books, and our calendar art all demonstrate, and that the United States has been, in many respects, a world leader in environmental protection. On the other hand, we have also been a nation of consummate nature destroyers, perhaps the most destructive ever; and again it has been the dream of living innocently in a bountiful Eden that is heavily responsible. Innocence, no matter how contrived or willful, can, after all, produce tragedy. So it has done in America. We have created out of our Edenic musings a peculiar sort of tragedy. Confident of having regained paradise, complacent and blissful in its midst, we have lost much of what we have most loved.

"Is it Un-Biocentric to Manage?"
A Response to Mike Seidman's Letter in Wild Earth, *Vol. 2 (30)*
William S. Alverson and Donald M. Waller

In an essay for *Natural History* conservation biologist Jared Diamond asks "Must we shoot deer to save nature?" He ponders the difficult choices faced by managers of natural areas, focusing on the Fontenelle Forest

Reserve near Omaha, Nebraska, a 1300 acre remnant of flood plain forest now isolated from other forest stands and surrounded by a matrix of suburban development and farm fields. Browsing by White-tailed Deer in the Fontenelle has eliminated regeneration in many tree species, reduced understory vegetation needed by many songbirds and butterflies, and, in all likelihood, eradicated several sensitive herbaceous species. Despite desires on the part of both the Forest managers and Diamond to avoid interfering in natural processes, Diamond reluctantly comes to the conclusion that shooting deer is necessary in this "tragic situation." The problem, Diamond argues, lies with context: this forest and many others have lost their large carnivores and are too small to sustain the natural ecological processes that might once have kept deer and other species in check. Super-normal densities of keystone browsers such as deer, in turn, have cascading effects that threaten diversity throughout the community.

Seidman disagrees with Diamond's assessment that management is necessary in such cases. He accuses Diamond of being an apologist for a "System" that invariably promotes the active management of natural and semi-natural habitats out of a desire to subjugate nature. Seidman deplores the compulsion evident in many land managers to actively manage their lands, arguing instead that leaving land alone will help us to shed our anthropocentrism. Perhaps inadvertently, he therefore sides with the animal rights activists who have opposed the shooting of deer so vigorously and effectively as to cancel deer control programs for the Fontenelle and elsewhere. Finally, he pleads for biocentric conservation biologists to speak out against what he sees as revisionist principles advanced in Diamond's essay.

As conservation biologists involved in a strenuous effort to convince the US Forest Service to adopt a minimal interference approach to managing substantial fractions of the two Wisconsin National Forests, we might seem appropriate candidates to accept Seidman's charge. For the past seven years, we have struggled through conversation, writing, negotiation, appeals, and litigation to convince the Forest Service that they could best and most cheaply meet federally mandated concerns to maintain biological diversity by abandoning active management for timber and game species in several large (50,000+ acre) "Diversity Maintenance Areas" (Solheim et al. 1987; Mlot 1992). Notably, Diamond wrote a strong letter in support of our attempts to create these DMAs, indicating clearly that he is not in favor of active management for its own sake. Finally, for the past three years we have conducted a large-scale research project into the key interaction that prompted Diamond's essay: the effects of deer browsing on sensitive plant species (Alverson et al. 1988). We can only agree with Mr. Seidman that land management policies of both public and private agencies show an overwhelming predilection toward manipulative management. Many public land managers shun the idea that leaving land alone might be good for it, considering it instead irresponsible or even immoral. This, of course, is nonsense. It is just as nonsensical, however, to assume that we should never actively manage land if our goal is to retain diversity or other natural values. Whenever possible, we should seek to reduce signs of human presence, reduce active management, and let natural forces reign. Yet in some instances we may be forced to intervene, even aggressively or frequently, if we wish to retain certain native elements of diversity.

Which threats justify intervention via active management and which should we simply accept? Is it appropriate to accept acidic precipitation on the Adirondacks and Appalachian wilderness areas? Should we watch silently as Paperbark (*Melaleuca*) and Brazilian Pepper trees (*Schinus*) invade the Everglades and displace native trees? Shall we refrain from burning prairie remnants in Iowa growing into Sweet Clover (*Melilotus*) and Box Elder (*Acer negundo*) because of the loss of a pre-settlement fire regime? Should we abandon cowbird control efforts and burning to retain young Jack Pine stands for Kirtland's Warbler in Oscoda County, Michigan? Should we attempt to reduce deer populations to increase the probability that moose now wandering into northern Wisconsin will become successful colonists (they now die because of a parasite shared with deer)?

In all these cases, we feel that biocentric managers have a responsibility to actively intervene to help compensate for the massive disruption that has already occurred in native ecological processes. While even purist biocentrists like Seidman would presumably favor actions to shield reserves from direct assaults like acid rain, it is difficult to argue that there is any real distinction among the threats mentioned above. No natural area in today's world is free of the "eternal external threat" posed by human disruption of local, regional, and global ecological processes (Janzen 1986). Edge-, area-, and isolation-sensitive plant and animal species are suffering in most parts of their range. External impacts, including perambulatory deer, penetrate to the core of small- to medium-sized reserves. While individuals may be unable to prevent global warming, conservationists do have the power and, correspondingly, the responsibility to protect reserves from threats they understand and can control. In such instances, *not* managing actively to control these threats is equivalent to accepting unwanted external interference. In this context, biologically informed and motivated efforts to minimize disruption of natural processes constitutes a far preferable form of interference.

In the case cited, stewards of the Fontenelle Forest face a restricted set of choices. They can: 1) allow overpopulation by deer to continue and label it "natural" (as Seidman seems to suggest); 2) drive the deer out and fence the area (very expensive but a realistic management option); 3) attempt to impose birth control on the deer (also expensive and usually unreliable); or 4) seek to control deer populations via hunting (effective and relatively cheap). All these alternatives have been thoroughly tried in dozens of forest preserves across the Northeast and Midwest and there is general agreement among natural areas managers which is preferable.

We would propose that the goal of biocentric land managers be to minimize interference in natural processes as far as possible. In general, larger nature reserves will require far less intervention and active management than small areas, one of many arguments in their favor. But if we are to retain diversity and the natural processes diversity depends upon, we will need to accept the obligation to manage smaller areas in accord with our best knowledge of what their constituent species require. Thus, we face the ironic (and, to some, unpalatable) situation that in order to achieve our goal of minimal interference, we will need to actively intervene in cases where we see a clear and present danger. In some cases, of course, we will not perceive a threat until it is too late. It is also likely that our efforts to actively manage natural areas will occasionally precipitate unintended consequences (one thinks of predators introduced onto islands to control mice and rats). Nevertheless, such errors do not justify inaction when clear and present dangers are recognized.

REFERENCES CITED

Alverson, W.S., S. Solheim, and D.M. Waller. 1988. Forests too deer: Edge effects in northern Wisconsin. *Conservation Biology* 2(4): 348–358.

Janzen, D.H. 1986. The eternal external threat. In M.E. Soulé, ed., *Conservation Biology: The Science of Scarcity and Diversity.* (Sunderland, Mass.: Sinauer Association), pp. 286–303.

Mlot, C. 1992. Botanists sue Forest Service to preserve biodiversity. *Science* 257: 1618–1619.

Solheim, S., W.S. Alverson, and D.M. Waller. 1987. Maintaining biotic diversity in National Forests: The necessity for large blocks of mature forest. *Endangered Species Technical Bulletin Reprint* 4(8): 1–3.

❧

From
"Ecology and
Revolutionary Thought"
Murray Bookchin

In almost every period since the Renaissance the development of revolutionary thought has been heavily influenced by a branch of science, often in conjunction with a school of philosophy.

Astronomy in the time of Copernicus and Galileo helped to change a sweeping movement of ideas from the medieval world, riddled by superstition, into one pervaded by a critical rationalism and openly naturalistic and humanistic in outlook. During the Enlightenment—the era that culminated in the French Revolution—this liberatory movement of ideas was reinforced by advances in mechanics and mathematics. The Victorian era was shaken to its very foundations by evolutionary theories in biology and anthropology, by Marx's contributions to political economy, and by Freudian psychology.

In our own time, we have seen the assimilation of these once-liberatory sciences by the established social order. Indeed, we have begun to regard science itself as an instrument of control over the thought processes and physical being of man. This distrust of science and of the scientific method is not without justification. "Many sensitive people, especially artists," observes Abraham Maslow, "are afraid that science besmirches and depresses, that it tears things apart rather than creating." What is perhaps equally important, modern science has lost its critical edge. Largely functional and instrumental in intent, the branches of science that once tore at the chains of man are now used to perpetuate and gild them. Even philosophy has yielded to instrumentalism and tends to be little more than a body of logical contrivances; it is the handmaiden of the computer rather than of the revolutionary.

There is one science, however, that may yet restore and even transcend the liberatory estate of the traditional sciences and philosophies. It passes rather loosely under the name "ecology"—a term coined by Haeckel a century ago to denote "the investigation of the total relations of the animal both to its inorganic and to its organic environment." At first glance, Haeckel's definition is innocuous enough; and ecology narrowly conceived of as one of the biological sciences, is often reduced to a variety of biometrics in which field workers focus on food chains and statistical studies of animal populations. There is an ecology of health that would hardly offend the sensibilities of the American Medical Association and a concept of social ecology that would conform to the most well-engineered notions of the New York City Planning Commission.

Broadly conceived of, however, ecology deals with the balance of nature. Inasmuch as nature includes man, the science basically deals with the harmonization of nature and man. The explosive implications of an ecological approach arise not only because ecology is intrinsically a critical science—critical on a scale that the most radical systems of political economy have failed to attain—but also because it is an integrative and reconstructive science. This integrative, reconstructive aspect of ecology, carried through to all its implications, leads directly into anarchic areas of social thought. For, in the final analysis, it is impossible to achieve a harmonization of man and nature without creating a human community that lives in a lasting balance with its natural environment.

WHAT WE ARE SEEING TODAY is a crisis in social ecology. Modern society, especially as we know it in the United States and Europe, is being organized around immense urban belts, a highly industrialized agriculture and, capping both, a swollen, bureaucratized, anonymous state apparatus. If we put all moral considerations aside for the moment and examine the physical structure of this society, what must necessarily impress us is the incredible logistical problems it is obliged to solve—problems of transportation of density, of supply (of raw materials, manufactured commodities, and foodstuffs), of economic and political organization, of industrial location, and so forth. The burden this type of urbanized and centralized society places on any continental area is enormous.

Diversity and Simplicity

The problem runs even deeper. The notion that man must dominate nature emerges directly from the domination of man by man. The patriarchal family planted the seed of domination in the nuclear relations of humanity; the classical split in the ancient world between spirit and reality—indeed, between mind and labor—nourished it; the anti-naturalist bias of Christianity tended to its growth. But it was not until organic community relations, feudal or peasant in form, dissolved into market relationships that the planet itself was reduced to a resource for exploitation. This centuries-long tendency finds its most exacerbating development in modern capitalism. Owing to its inherently competitive nature, bourgeois society not only pits humans against each other, it also pits the mass of humanity against the natural world. Just as men are converted into commodities, so every aspect of nature is converted into a commodity, a resource to be manufactured and merchandised wantonly. The liberal euphemisms for the processes involved are "growth," "industrial society," and "urban blight." By whatever language they are described, the phenomena have their roots in the domination of man by man.

The phrase "consumer society" complements the description of the present social order as an "industrial society." Needs are tailored by the mass media to create a public demand for utterly useless commodities, each carefully engineered to deteriorate after a predetermined period of time. The plundering of the human spirit by the marketplace is paralleled by the plundering of the earth by capital. (The liberal identification is a metaphor that neutralizes the social thrust of the ecological crisis.)

ECOLOGISTS ARE OFTEN ASKED, rather tauntingly, to locate with scientific exactness the ecological breaking point of nature—the point at which the natural world will cave in on man. This is equivalent to asking a psychiatrist for the precise moment when a neurotic will become a nonfunctional psychotic. No such answer is ever likely to be available. But the ecologist can supply a strategic insight into the directions man seems to be following as a result of his split with the natural world.

From the standpoint of ecology, man is dangerously oversimplifying his environment. The modern city represents a regressive encroachment of the synthetic on the natural, of the inorganic (concrete, metals, and glass) on the organic, of crude, elemental stimuli on variegated, wide-ranging ones. The vast urban belts now developing in industrialized areas of the world are not only grossly offensive to the eye and the ear, they are chronically smog-ridden, noisy, and virtually immobilized by congestion.

The process of simplifying man's environment and rendering it increasingly elemental and crude has a cultural as well as a physical dimension. The need to manipulate immense urban populations—to transport, feed, employ, educate, and somehow entertain millions of densely concentrated people—leads to a crucial decline in civic and social standards. A mass concept of human relations—totalitarian, centralistic, and regimented in orientation—tends to dominate the more individuated concepts of the past. Bureaucratic techniques of social management tend to replace humanistic approaches. All that is spontaneous, creative, and individuated is circumscribed by the standardized, the regulated, and the massified. The space of the individual is steadily narrowed by restrictions imposed upon him by a faceless, impersonal social apparatus. Any recognition of unique personal qualities is increasingly surrendered to the manipulation of the lowest common denominator of the mass. A quantitative, statistical approach, a beehive manner of dealing with man, tends to triumph over the precious individualized and qualitative approach which places the strongest emphasis on personal uniqueness, free expression, and cultural complexity.

The same regressive simplification of the environment occurs in modern agriculture. The manipulated people in modern cities must be fed, and to feed them involves an extension of industrial farming. Food plants must be cultivated in a manner that allows for a high degree of mechanization—not to reduce human toil but to increase productivity and efficiency, to maximize investments, and to exploit the biosphere. Accordingly, the terrain must be reduced to a flat plain—to a factory floor, if you will—and natural variations in topography must be diminished as much as possible. Plant growth must be closely regulated to meet the tight schedules of food-processing factories. Plowing, soil fertilization, sowing, and harvesting must be handled on a mass scale, often in total disregard of the natural ecology of an area. Large areas of the land must be used to cultivate a single crop—a form of plantation agriculture that not only lends itself to mechanization but also to pest infestation. A single crop is the ideal environment for the proliferation of pest species. Finally, chemical agents must be used lavishly to deal with the problems created by insects, weeds, and plant diseases, to regulate crop production, and to maximize soil exploitation. The real symbol of modern agriculture is not the sickle (or, for that matter, the tractor), but the airplane. The modern food cultivator is represented not by the peasant, the yeoman, or even the agronomist—men who could be expected to have an intimate relationship with the unique qualities of the land on which they grow crops—but the pilot or chemist, for whom soil is a mere resource, an inorganic raw material.

The simplification process is carried still further by an exaggerated regional (indeed, national) division of labor. Immense areas of the planet are increasingly reserved for specific industrial tasks or reduced to depots for raw materials. Others are turned into centers of urban population, largely occupied with commerce and trade. Cities and regions (in fact, countries and continents) are specifically identified with special products—Pittsburgh, Cleveland, and Youngstown with steel, New York with finance, Bolivia with tin, Arabia with oil, Europe and the U.S. with industrial goods, and the rest of the world with raw materials of one kind or another. The complex ecosystems which make up the regions of a continent are submerged by an organization of entire nations into economically rationalized entities, each a way station in a vast industrial belt-system, global in its dimensions. It is only a matter of time before the most attractive areas of the countryside succumb to the concrete mixer, just as most of the Eastern seashore areas of the United States have already succumbed to subdivisions and bungalows. What will remain in the way of natural beauty will be debased by trailer lots, canvas slums, "scenic" highways, motels, food stalls, and the oil slicks of motor boats.

The point is that man is undoing the work of organic evolution. By creating vast urban agglomerations of concrete, metal, and glass, by overriding and undermining the complex, subtly organized ecosystems that constitute local differences in the natural world—in short, by replacing a highly complex, organic environment with a simplified, inorganic one—man is disassembling the biotic pyramid that supported humanity for countless millennia. In the course of replacing the complex ecological relationships, on which all advanced living things depend, for more elementary relationships, man is steadily restoring the biosphere to a stage which will be able to support only simpler forms of life. If this great reversal of the evolutionary process continues, it is by no means fanciful to suppose that the preconditions for higher forms of life will be irreparably destroyed and the earth will become incapable of supporting man himself.

Ecology derives its critical edge not only from the fact that it alone, among all the sciences, presents this awesome message to humanity, but also because it presents this message in a new social dimension. From an ecological viewpoint, the reversal of organic evolution is the result of appalling contradictions between town and country, state and community, industry and husbandry, mass manufacture and craftsmanship, centralism and regionalism, the bureaucratic scale and human scale.

ॐ

From
"Man's Place in Nature"
Joseph Wood Krutch

To what extent then should man, to what extent dare he, renounce nature; take over the management of the earth he lives on; and use it exclusively for what he sometimes regards as his higher purposes?

Extremists give and have always given extreme answers. Let us, say some, "return to nature," lead the simple life, try to become again that figment of the romantic imagination, "the noble savage." Henry David Thoreau, the greatest of American "nature lovers," is sometimes accused of having advocated just that. But he did not do so; he advocated only that we should live more simply and more aware of the earth, which, he said with characteristic exaggeration, "is more wonderful than it is convenient; more beautiful than it is useful; it is more to be admired and enjoyed than used."

Others suggest a different extreme. They talk about "the biosphere" (loosely, that which has been here defined as the natural world) as contrasted with "the noosphere" (translated as that portion of the earth upon which man has imposed his own will so successfully that whatever conditions prevail there do so because of his will). It appears that civilization, according to this notion, is to be completed only when the noosphere is the whole earth and the biosphere is completely subordinated to the human will.

Within the last one hundred years we have approached faster and closer to that condition than in all the preceding centuries of civilization. But would man, whose roots go so deep into nature, be happy should he achieve such a situation?

Certainly he would become a creature very different from what he is, and the experience of living would be equally different from what it has always been. He would, indeed, have justified his boast that he can "conquer nature," but he would also have destroyed it. He would have used every spot of earth for homes, factories, and farms, or perhaps got rid of farms entirely, because by then he could synthesize food in the laboratory. But he would have no different companions in the adventures of living. The emotions which have inspired much of all poetry, music, and art would no longer be comprehensible. He would have all his dealings with things he alone has made. Would we then be, as some would imagine, men like gods? Or would we be only men like ants?

That we would not be satisfied with such a world is sufficiently evidenced by the fact that, to date at least, few do not want their country house, their country vacation, their camping or their fishing trip—even their seat in the park and their visit to the zoo. We need some contact with the things we spring from. We need nature at least as a part of the context of our lives. Though we are not satisfied with nature, neither are we happy without her. Without cities we cannot be civilized. Without nature, without wilderness even, we are compelled to renounce an important part of our heritage.

The late Aldo Leopold, who spent his life in forestry and conservation, once wrote: "For us of the minority, the opportunity to see geese is more important than television, and the chance to see a pasqueflower is a right as inalienable as free speech."

Many of us who share this conviction came to it only gradually. On some summer vacation or some country weekend we realize that what we are experiencing is more than merely a relief from the pressures of city life; that we have not merely escaped *from* something but also *into* something; that we have joined the greatest of all communities, which is not that of men alone but of everything which shares with us the great adventure of being alive.

This sense, mystical though it may seem, is no delusion. Throughout history some have felt it and many have found an explanation of it in their conviction that it arises out of the fact that all things owe their gift of life to God. But there is no reason why the most rationalistic of evolutionists should not find it equally inevitable. If man is only the most recent and the most complex of nature's children, then he must feel his kinship with them. If even his highest powers of consciousness, intellect, and conscience were evolved from simpler forms of the same realities, then his kinship with those who took the earlier steps is real and compelling. If nature produced him, and if she may someday produce something far less imperfect, then he may well hesitate to declare that she has done all she can for him and that henceforth he will renounce her to direct his own destiny.

In some ways man may seem wiser than she is, but it is not certain that he is wiser in all ways. He dare not trust her blindly, but neither does he dare turn his back upon her. He is in danger of relying too exclusively upon his own thoughts, to the entire neglect of her instincts; upon the dead machine he creates, while disregarding the living things of whose adventure he is a part.

We have heard much about "our natural resources" and of the necessity for conserving them, but these "resources" are not merely materially useful. They are also a great reservoir of the life from which we evolved, and they have both consolation to offer and lessons to teach which are not alone those the biologist strives to learn. In their presence many of us experience a lifting of the heart for which mere fresh air and sunshine is not sufficient to account. We feel surging up in us the exuberant, vital urge which has kept evolution going but which tends to falter amid the complexities of a too civilized life. In our rise to the human state we have lost something, despite all we have gained.

Is it merely a sentimental delusion, a "pathetic fallacy," to think that one sees in the animal a capacity for joy which man himself is tending to lose? We have invented exercise, recreation, pleasure, amusement, and the rest. To "have fun" is a desire often expressed by those who live in this age of anxiety and most of us have at times actually "had fun." But recreation, pleasure, amusement, fun, and all the rest are poor substitutes for joy; and joy, I am convinced, has its roots in something from which civilization tends to cut us off.

Are at least some animals capable of teaching us this lesson of joy? Some biologists—but by no means all and by no means the best—deny categorically that animals feel it. The gift for real happiness and joy is not always proportionate to intelligence, as we understand it, even among the animals. As Professor N.J. Berrill has put it, "To be a bird is to be alive more intensively than any other living creature . . . [Birds] live in a world that is always the present, mostly full of joy." Similarly Sir Julian Huxley, no mere sentimental nature lover, wrote after watching the love play of herons: "I can only say that it seemed to bring such a pitch of emotion that I could have wished to be a heron that I might experience it."

This does not mean that Sir Julian would desire, any more than you or I, to be permanently a bird. Perhaps some capacity for joy has been, must be, and should be sacrificed to other capacities. But some awareness of the world outside of man must exist if one is to experience the happiness and solace which some of us find in an awareness of nature and in our love for her manifestations.

Those who have never found either joy or solace in nature might begin by looking not for the *joy they can get,* but for the *joy that is there* amid those portions of the earth man has not yet entirely pre-empted for his own use. And perhaps when they have become aware of joy in other creatures they will achieve joy themselves, by sharing in it.

❦

"Antelope"
Charles Bowden

The men stand on the runway and talk of Nam, Beirut, and Grenada and wonder about the possibilities of Central America. Two fixed-wing aircraft and three choppers wait on the cool cement this morning. The men are pilots and they savor war news—old wars, recent wars, future wars. Some of the men are in the Air Force and some are now out but the taste of action, the chance to be there, to be where it is really happening, tugs at them. They believe in American might and they are wounded by each American humiliation. Iran? Oh yes, Iran, they say. That must not happen again.

The tiny runway lies a few miles out of Ajo, a dying Arizona copper town about fifty miles north of the Mexican line. The slabs were poured in World War II to train pilots for Europe and Asia. When that war fell away, the installation went to ruin. Weeds peek now between the cracks in the concrete. The old officers' quarters just to the east has become a country club for the small community's few copper executives.

I lie on the runway and listen to the talk. I have never been in any army or any war but I am not immune to these feelings. The Air Force machines whip up a martial ardor in the men, military and civilian alike. Words stenciled across the chopper act as cue cards: EXIT RELEASE; PULL SLIDE; BACK ASST.; OPERATING LIMITS; DANGER.

We are on a set and the set makes us all actors.

Off to the west, on Childs Mountain, a new military outpost pokes up against the horizon. Here millions of dollars of computers, radio equipment, and high-tech witchcraft monitors all the firing on the gunnery range up to the north and west. There is talk of lasers teasing the targets, of drones being towed like cows for practice shooting. Out there the desert is littered with brass and people sneak in to salvage it. Sometimes they get blown up. Sometimes they die of thirst.

We wait to go into that shooting gallery. The day is scheduled to be a cold day—one in which the Air Force will refrain from clotting the sky with hot rounds.

We will not trigger cannons. We prefer nets. And we don't stalk the enemies of the state but pursue a more slippery prey, a life force 10 to 25 million years old.

They have always been Americans. All the fossils and all the living forms on Earth come from North America. Once great herds stretched from southern Canada down into Mexico. The subspecies roaming west of Ajo (*Antilocapra americana sonoriensis*) was identified only in 1945, based on an adult female skin and two skulls. The key specimen was killed December 11, 1932. From these bones Americans learned of another American.

We barely know our quarry. We do not know how many there are. We do not know where they go. We do not know how they organize themselves. We do not even know if they drink water. And we cannot imagine what they think.

They are commonly called Sonoran pronghorn. They occupy a small patch of southwestern Arizona and northwestern Sonora. Maybe three or four hundred survive on the surface of the earth, perhaps seventy-five to one hundred, we hope, romp on the American side of the fence. Once they roamed across the Colorado River deep into California and for more than one hundred miles east of Ajo to the edges of Tucson and Phoenix. The ways of man and man's beasts have driven them deep into the dry heat of the desert and here they linger, an echo of a world largely slaughtered long ago.

Mexico has not permitted a legal hunt since 1922 but this law is sometimes winked at. Twenty years ago, four heads from Mexico were seized in a Tucson taxidermy shop. Today, in Sonora, the killing continues disguised with weird government permits and thick bundles of cash slid under the table. In the world of trophy hunting, there are always a few sports who go off the rails and become strange Noahs filling their death ships with one of each species.

On the American side, the antelope now roam over about 2.5 million acres. This is the place where moisture is a dream. At Ajo, the average annual rainfall is 9.1 inches. One hundred and thirty miles to the west, at Yuma, 3.38 inches. And some years it forgets to rain.

The pronghorn blend so well into this minimal landscape that they are almost invisible. I have spoken to game officials who patrol this turf and who have gone for weeks and weeks without sighting one. A recent survey turned up only thirteen in three days of flying.

They survey here because the military needs a place to train for war and this kind of work keeps just about everyone else out—the poachers, the cattle, the hurly burly of twentieth century ways and means. The land in the military gunsights is part of the Cabeza Prieta National Wildlife Refuge plus the Luke-Williams Gunnery range. Naturally, the Air Force did not originally consider whether rare antelope bounced along on their playing field—or anything else for that matter. But now that the subspecies has been brought to their attention, they are willing to make a good show of it. Luke is bucking to win the Air Force conservation award and tending to endangered antelope looks like a good way.

So they've loaned pilots and aircraft for this hunt. Along with personnel from the U.S. Fish and Wildlife Service and the Arizona Game and Fish Department, the Air Force men will join in an effort to capture, examine, and radio collar at least five antelope. It is hoped that something will be learned that will help pull the small bands back from the abyss of extinction.

At 7 A.M. the two planes take off to scout up some antelope. The choppers and the bulk of us wait around the runway for the call, a wait that could go on and on and on. The hunt has all sorts of little quirks: the antelope cannot be pursued if the air temperature exceeds 80 degrees because they might die from heat stress; they cannot be held captive for more than five minutes for the same reason.

Because we are so ignorant of the antelopes' habits, they cannot be baited into a trap—we have no idea what they like to eat. A new technique, firing a net from a gun powered by a .308 cartridge, has come into recent vogue (no wildlife has yet perished by this method) and this is the weapon of choice this morning.

While the antelope are rare, the bureaucratic agreement on this old runway is even rarer. For several years, the various agencies have haggled over details, money, and tactics. It is a wonder the antelope managed to survive long enough to be chased by helicopters and manhandled by human beings.

One federal official has flown in from Albuquerque, several state experts have come from Phoenix, and a small herd of television and print reporters have journeyed from Tucson. None of us came here with terribly clean hands.

The Air Force rules lay down all kinds of restrictions out in the desert Refuge: no plane should normally fly below 3000 feet or ever below 500 feet. I've seen them 100 feet off the ground and one pilot boasted to a friend of mine that he cracked the sound barrier 25 feet from the ground. There are areas where no firing should take place, like waterholes. I've found spent rockets lying around waterholes favored by bighorn sheep. But then, we're hardly likely to recruit men good at flying fighter planes who tend to obey rules. Anyway, the antelope seem to adjust to the jets— they've been sighted feeding with aircraft at 500 feet and have not even raise their heads.

The people from the state Game and Fish Department have a checkered past with antelope—one of the great failures of a general wildlife revival in the state. Antelope can't seem to master jumping fences and as the state goes to more barbed wire and little ranchette estates, remaining sanctuaries for the species are disappearing. The Department responds by haggling a little here and there and periodically shooting coyotes from airplanes to cut down predation of the fawn crop.

The U.S. Fish and Wildlife people have a few bad moments to live with too. As lackeys of the state cattle interests they almost single-handedly exterminated the wolf (last den with pups found in 1943 and slaughtered) and grizzly bear (last sighted in 1936).

I myself have become almost a functionary for the Air Force in this antelope business, merrily cranking out newspaper stories lauding their efforts in the hope they will accept the delusion and make a major effort to safeguard the beasts.

Meanwhile the antelope have been run over by cars on the highway, poached here and there, drowned in irrigation ditches. And survived. So here we are armed with big machines and science to lend them a helping hand.

At 9 A.M. the scouts call in that antelope have been sighted and we all race for our choppers like Green Berets. We cross the black mountains and cruise the creosote flats and within twenty minutes the band is found. Nine antelope romp along at fifty miles an hour. Two choppers stay one hundred feet off the desert floor and the capture chopper drops down to about eight feet.

The machine dogs the animals and they wheel and turn and charge and spin, trying to shake the strange predator. One man hangs out the open door clutching the gun. After three minutes of harassing the band to wear them out, the man fires and an antelope goes down trapped in the maze of an orange net. All three craft land and we race toward our quarry.

Huge spikes drink the beast's blood, a thermometer is rammed up its ass, and a white radio collar snapped on the neck. The antelope is blindfolded to calm it and this spares us the wild look in its eyes. The animal writhes and the body temperature, normally 98 or 99, jumps to 104. We all jockey in and touch. We cannot resist.

There are four or five billion of us, and a couple of dozen of them. Everyone must touch.

For years there has been speculation that the pronghorn are mute. They are not.

A deep rattling moan escapes the animal and the warm desert air suddenly chills. No one can mistake the message of the sound, the grieving powering of it.

A kind of collision between cultures has taken place. Huge machines that fly at eighty miles per hour and drink more than seventy gallons an hour have snared an organism that has raced at fifty miles per hour for millions of years.

The sound keeps coming and coming. We have violated some deep important thing and we cannot doubt this fact.

From
The Tender Carnivore and the Sacred Game
Paul Shepard

Making Wild Sheep Tame

The dawn of civilization, associated with the first agriculture, is generally seen as a great sunrise before which men lived in a mental and social twilight, waiting, straining to become fully human. We see those vague predecessors as incomplete, with a few crude tools, living days of fear and monotony, nights of terror and discomfort, with a short, brutish existence as the only reward for the struggle to survive.

Today's myth of progress and gospel of radical change, orientation to tomorrow and frantic exchange of old things for new are modern only in terms of the whole human span that preceded them. Though we may picture ourselves as very unlike old world peasants it is in the agrarian mind that modern life begins.

Nobody knows for certain how or why agriculture began. How man began the tending of plants and animals may remain a secret forever, but the epidemic of acquisitive proprietorship and territorial aggrandizement that resulted from this development is all too apparent. From what evidence there is it was neither a worldwide event nor a single event, but a shifting mixture of hunting, fishing, and planting, at first in a rather limited geographical area from Turkey to the Caspian basin and south to the Red Sea and Palestine and, later, parallel expansions from centers in equatorial Asia and America.

Despite many specific domestications and forms of early agriculture in other parts of the world, particularly Southeast Asia and Central America, the set of techniques and the set of mind seem first to have taken form at the eastern end of the Mediterranean Sea and to have spread from there by cultural contact. In time, Asia and America became centers of domestication as agriculture spread around the globe.

The hunting and gathering peoples who preceded the planters did not live in an unchanging world, nor were they themselves alike everywhere or at different times. Diversity existed because of human adaptations to local environmental differences. The climate, species of plants and animals, types of available food, and the manmade combinations of tools and ideas had been changing for a million years. Men had expanded their range, shifted with the slow tides of glacial ice, learned the ways of other living creatures, and achieved a rich humanity long before metals, pots, wheels, kings, and theocracies appeared.

Because of the limited archaeological evidence, it was still thought until recently that domestication appeared abruptly, as an inexplicable breakthrough that transformed human life about ten thousand years ago. Now it is clear that its sudden appearance in archaeological diggings was a local accident. Between outright gathering and full food production there were stages of food-collecting and incipient domestication, life ways mixing hunting and gathering in different proportions, and finally the culling or artificial hunting of animals and collecting from extended garden plants that were more or less constantly protected and probably genetically altered by men.

The story of some cave-man genius bringing home a baby wild sheep to raise or capturing a cub-wolf from its den, or realizing in some bold flash of intuition how to grow tomatoes, thus inventing in one mighty stroke a new way of life that was thereafter imitated by the genius's friends and descendants, is nonsense; it is part of the same civilized myth that would have us believe that in some miraculous way the farmer discovered agriculture and thus raised himself above his predecessors.

A sequence of events east of the Mediterranean Sea some twenty thousand years ago may have led by uncertain steps toward agriculture. Whether these stresses and deformations of human society were the consequence of the climate and the glaciers of the last ice age has been argued inconclusively for half a century. The great wetness of the time has suggested to some that men who had previously chased big game were impeded by the waterways and swamps, that hunting faltered as large mammals became locally extinct, that fishing may have seduced men into a sedentary life, which made for a different kind of attention to the plants and earth around them. It has been suggested that after the wet came the dry; that the principal wild herd animals remaining were aurochs (the wild cattle) and the ancestors of domestic goats and sheep; and that grasses flourished on the slopes so abundantly that their seeds became an increasingly available food for people. If the use of cereals was the crux of man's pre-history, then it was inseparable from the use of fire for cooking, making them edible in large amounts.

During this stage of deteriorating cynegetics (that is, hunting and its culture) local tribes apparently tried a great variety of foods that earlier had been rejected by the reindeer hunters who preceded them. Archaeological records have shown signs of crisis during the four thousand years preceding the first farming communities. It was a time of intense food experimentation in which acorns, nuts, seeds, snails, clams, fish and other aquatic animals were gathered. The mastery of stone tool-making faded, but there were new utensils of wood, leather, and bone, evidence of swimming and the use of boats, flint sickles, the pestle and mortar, and the bow and arrow. These innovations by seed-gleaners, shore-scavengers, and sheep-followers in the Tigris-Euphrates watershed and on the slopes of the Zagros and Palestinian mountains preceded the earliest concrete signs of domestication—the bones of livestock and seed types found only in association with man.

This was a time of trial and difficulty for the societies and cultures that had created an art and religion that had endured for twenty thousand years over much of Europe and Asia, yet it favored experiment and goaded men into versatility. By the time the climate warmed around the Mediterranean, some twelve thousand years ago, sedentary, seed-conscious people were munching local mutton in villages. For them the bison, the woolly mammoth and the rhino, the wild horse, cave bear, and reindeer, had all but passed from memory.

The lands they occupied may have looked very different from the more northerly tundras and steppes or the more southerly savannas where hunters still flourished. The rolling, open, upland terrain was flecked with patches of woodland and streams separated by grassy swards. Roaming herds of wild sheep and goats gradually became habituated to people, tolerating human presence even within the "flight-distance" at which wild animals normally flee. They were an increasingly easy kill for men whose durability, strength, and cunning had been honed through generations by their hunting forebears. But compared to their ancestors' grand chases and spearing of wild horses, plugging goats with arrows must have been a profound though perhaps unconscious disappointment.

The virtual collapse of hunting and gathering, the central activity of the ancient culture, would surely have affected the very heart of human existence. The great mystery of domestication is therefore not so much how men achieved control of plants and animals, but how human consciousness was reorganized when the cynegetic life was shattered—that is, the mental, social, and ecological complex based on hunting. All major human characteristics—size, metabolism, sexual and reproductive behavior, intuition, intelligence—had come into existence and were oriented to the hunting life. How, for example, did the male prerogative shift from the chase to the harvest of plants? Even though men at all times have continued to hunt with great elan and pride, just as women have continued to gather and to be the center of the household, it is astonishing that such a shift in vocation took place.

In the course of a few thousand years men assumed control over the harvest, however much it was ritualized in female symbols. Throughout history agriculture has been represented in feminine terms and images. Even so, men generally dominate the political order in such societies—as among Biblical peoples. Perhaps this is because, from the earliest times of farming, a major purpose of man's pastures and fields was the production of meat from grazing animals and the harvest of meat was the ancient domain of the male as a hunter. Or did control of surplus grain in some way acquire the prestige and potential power traditionally associated with the hunt and the hunter's honor in distributing the parts of the kill?

Driving the bezoar goat and urial sheep, however it may have been compensated in social status, must have seemed, as the animals became more and more tame, less worthy of a man's life. Cut off from hunting reindeer, horse, and elephant, men lost both the models and the means by which personal integrity was achieved and measured within the group by peaceful means. They found a substitute in the biggest and most dangerous potential prey remaining—men themselves.

The collapse of an ecology that kept men scarce and attuned to the mystery and diversity of all life led as though by some devilish Fall to the hunting and herding of man by man, to the hoarding of grain and the secularizing of all space. To defend his fields a farmer needed many kinsmen: sons, and co-defenders, and co-fighters, and ultimately brother ideologists.

There are many physical and environmental aspects of domestication that are still unknown. Why in more than two million years of hunting, during which climates and animals had waxed and waned, had it not happened before? What was the true order of such interrelated events as preparing grain foods by cooking, monitoring of herds, driving and butchering based on the calendar, planting and tending seeds, stockading and breeding captive animals? Even the function or purpose of the control of different species of plants and animals is imperfectly known. Were animals first kept in corrals for religious reasons or did the religious and food motives develop simultaneously?

It now appears that what had been so long thought to be the "beginning" is the climax of events that took place over a span of time three times as great as the Christian era. The great moments in the origin of agriculture were, instead, a blend or mix of activities, during which men increased their reliance on seed foods, set fires that modified the vegetation, and hunted an ever-narrowing range of medium-sized mammals, which they protected from other predators. The seeds of the ancestral wheat and barley plants were simply gathered long before they were deliberately sown.

It was once thought that man first became sedentary and then noticed luscious pods and fruits that sprouted from his own dung and garbage heaps; or that men already sedentary captured animals because there was at last a place to keep them. It is now known that men built and used efficient shelters, developed language and complex cultures, made beautiful tools, and kept fires in what are now Turkey, Czechoslovakia, France, Russia, Hungary, and probably elsewhere hundreds of centuries before agriculture.

Other first steps in the development of agriculture have been postulated, but none can explain the events between fourteen thousand and twelve thousand years ago. Even pot-making, the celebrated art of the food-production revolution, had long been anticipated insofar as containers were needed and made by people who collected but did not plant, store, or barter food. It has been said in half-seriousness that as man tamed the dog the dog tamed man, but the dog may not even have been among the first domesticated organisms: the evidence is inconclusive because of the similarity of wild and tame canid bones at archaeological sites. A reliable chronology of the earliest presence of each of the hundreds of domestic plants and animals is slowly emerging from carbon-14 and other dating methods. Rather than explaining the beginnings of agriculture, these are after the fact and do not reveal the crucial turn at agriculture's beginnings.

What must surely have preceded farming was a shift in style and in man's sense of his place in the world; a shift whereby man would presume to own the world and wild organisms would be screened for those having a certain infantile, trusting placidity that could be nurtured and increased in captivity. Long before some degenerate auroch or wild cow was hitched to a set of stone wheels in Egypt, the destiny of the planet was altered. A group of deprived hunters, caught in a geographic and biological crisis, took up crayfish-stomping and seed-gleaning—activities that had not occupied the full attention of their ancestors for millions of years, since the earliest genus of man, *Ramapithecus*, ambled about pond and prairie edges and the father of modern men, *Australopithecus*, scavenged, sought small game, and snatched crustaceans in the shallows of ancient African lakes.

Husbandry, a Failure of Biological Style

It is hard to speak of domestic animals as failures because we are so fond of them. To us they are fellow beings, whereas we regard wild creatures as curiosities or shadows whose wills oppose our own. To denounce farming and rural life, so relatively serene at a time of urban crisis, seems to flout the last landscape of solace and respite. But in my view the urban crisis is a direct consequence of the food-producing revolution. In a sense, although farmers domesticated particular varieties of plants and animals, the farm domesticated the habitat. For whom was food mass-produced and food surpluses stored if not for the town?

"Domestication" means much more than the dictionary definition, "to become a member of the household." Individual wild creatures brought into the house, no matter how much they are loved or how long they are kept, do not become domesticated. If they live they may not thrive so well as in the wild; if they thrive they still may not reproduce or generate a lineage; and if they breed, the offspring may still prefer to go free if they can escape. There is an inborn difference between domesticates and all other animals.

"To domesticate" means to change genetically, to alter a group of organisms so that their behavior and appearance are quite different from their wild relatives, and these changes are transmitted to their offspring. By selecting parents, culling undesirables, inbreeding and crossbreeding, man uses the same processes that operate in nature.

Each gene in an individual organism acts in the context of many other genes. Hence the genetic changes resulting from domestication may affect the whole creature, its appearance, behavior, and physiology. The temperament and personality of domestic animals are not only more placid than their wild counterparts, but also more flaccid—that is, there is somehow less definition. Of course there is nothing placid about an angry bull or a mean watchdog, but their mothers were tractable, and once an organism has been stripped of its wildness it can be freaked in any direction the breeder wishes. It may be made fierce without being truly wild. The latter implies an ecological niche from which the domesticated animal has been removed. Niches are hard taskmasters. Escape from them is not freedom but loss of direction. Man substitutes controlled breeding for natural selection; animals are selected for special traits like milk production or passivity, at the expense of over-all fitness and naturewide relationships. All populations are composed of individuals who differ from one another. Among wild animals, the diversity is constantly pared at its fringes. Reproductive success and survival are best for individuals of a certain type. In this way, natural selection is a stabilizing pressure, shaping populations into distinctly different and recognizable species. This pressure does not exclude genetic variation. Indeed, the appearance and behavior of a wild species hold true to type in spite of genetic variation. Apparent uniformity masks genetic difference. When natural selection is removed much of that hidden variation emerges and the population is flooded with external diversity.

Our subjective experience of this is in terms of individuality, and the concept of individuality in our society carries such a strong emotional force as well as political overtones that individualizing as a by-product of domestication may not easily be seen as undesirable. Though domestication broadens the diversity of forms—that is, increases visible polymorphism—it undermines the crisp demarcations that separate wild species and cripples our recognition of the species as a group. Knowing only domestic animals dulls our understanding of the way in which unity and discontinuity occur as patterns in nature, and substitutes an attention to individuals and breeds. The wide variety of size, color, form, and use of domestic horses, for example, blurs the distinctions among different species of *Equus* that once were constant and meaningful.

It is important to know, when any two organisms are compared, how they are related. However trivial that distinction may seem at first, its triviality simply signifies the poverty of biology in modern philosophy. With domestication, arbitrary re-establishment of inconsequential groupings and relationships damaged the perceptual powers of mankind. If evolutionary human ecology had only one lesson it would be that the development of human intelligence is linked to man's conscious exploration of the species system in nature. But this lost-sensibility aspect carries us away from the biological consequences of domestication.

The glandular and anatomical alterations of animal domestication are fairly well known. Consider the white rat, whose history has been comparatively well documented since it was created from the wild brown rat in the middle of the nineteenth century. In breeding for ease of keeping and uniformity, a variety of related and inadvertent changes have occurred. The tamer, more tractable, less aggressive, more fecund white rat, with its early gonadal development, less active thyroid, and smaller adrenals, is cursed with greater susceptibility to stress, fatigue, and disease and has less intelligence than its wild relative. Many of the deleterious changes have been unavoidable side effects, because of the interplay of gene action, or because genes favored by the breeder are closely linked to undesirable genes on the chromosomes.

The strong, firm style of the wild animal is due to a mix of genes that work well together—in other words, it has a stable epigenetic system. The similarity of individuals is shielded by this system against disruption by mutation. In addition, certain chromosomal aberrations may serve to keep blocks of genes together in the wild form; that is, they tend to reduce mixing or recombination in later generations. In domestication the breeder breaks up these blocks, allowing new combinations to appear in the offspring, which would render most of them unfit in nature. Some of these will be especially desirable to him, others simply monsters. Once a cluster is broken up by man-controlled breeding producing genetic "goofies" that are protected from the rigors of the wild, new sets of captivity types can then be winnowed from them on the farm or in the laboratory. Even in zoos, where the majority of wild animals soon die, the captives that survive are those which, because of their genetic difference, are least exacting about territory, least subtle about social signals and cues, least precise in behavioral discrimination, the loss of mates and companions, and fear about human ubiquity. All domestic animals are highly social, but their social relations are degraded and generalized, just as their physiology is radically altered. They have been bred for readiness to accept human control. "Releasers"—those signals from others of their kind which trigger complex behavior sequences—are lost, along with genetically regulated responses. For them the world grows simpler.

Ritual behavior becomes abbreviated. Symbolic fighting to settle conflicts peacefully is less frequent among domestics than among wild species. Mating patterns lose their elaborate timing as segments are lost. Hormonal changes, such as a decrease in adreno-cortical steroids, lead to submissiveness. The reproductive systems of the kept animals lose their fine tuning to the season and to display postures, which in nature are a tightly woven sequence of steps from courtship through parental behavior. Differences between male and female—the secondary sexual characteristics—are diluted. The animals become crude pawns in the farmer's breeding game, shorn of finesse and the exquisite detail so characteristic of wild forms. The animal departs from the hard-won species type. For man the animal ceases to be an adequate representation of a natural life form. Its debased behavior and appearance mislead us and miseducate us in fundamental perceptions of the rhythms of continuity and discontinuity, and of the specific patterns of the multiplicity of nature.

Interpretations of this debauched ecology were formulated for civilization by its "educated" members. Effete dabblers from the city looked over the barnyard fence at the broken creatures wallowing and copulating in their own dung, and the concept of the bestial brute with untrammeled appetites was born. This was the model for "the animal" in philosophy, "the natural" from which men, understandably, yearned for transcendent release.

Among animals, suitable candidates for domestication are social, herd-oriented, leader- or dominance-recognizing forms. Their response to their own species (possible sex partners) and their own habitat is more a matter of learning and less of fixed responses to fixed signals. Husbandry seeks out and exploits three characteristics of these animals: the tendency of the young to follow whoever is caring for it by imprinting—the process of irreversible attachment; the gradualness of the transition from nursing to eating; and the way in which different social relations may be mediated by different senses. For example, mother–daughter nurture relationships may be based on an imprinted taste. A Scottish milkmaid lets the cow lick her bloodied hands (as well as the calf) at birth, and thereafter the cow will "let down"—give milk—for the milkmaid and the calf, but only for them.

Inborn metabolic errors condemn wild animals to swift destruction. In captivity such cripples are sometimes not only protected but prized.

These flaws ("hypertrophies") in growth result in the production of extra meat, wool, silk, eggs, and milk. All such freaks carry a burden of genetic weakness. The nurture of these weaklings is a large part of modern animal science, which may be defined as the systematic creation of animal deformities, anomalies, and monsters and the practice of keeping them alive.

Another mutant trait common to domestics is excessively delayed maturity and sexual precocity combined with rapid growth. In culling out the irascible and stubborn individuals, the hard, mature, lean line is sacrificed for animals with submissive and infantile responses. Individuals maturing at slower rates are favored. Cows and horses have long-enduring mother–child relationships just as primates do. By exploiting this relationship, new social interdependencies can be created. Infantile animals are less attached to their own kind and readily join other barnyard animals or the human household. Children are eager to adopt them as "people" and adult humans are attracted by their helpless appeal and immature faces—for juvenile qualities are as apparent in face and body as in behavior. The effect of all this is that domestic breeds are creatures who never grow up in spite of their sexual precocity.

The protected environment of domestic animals cushions them from the sculpturing forces of nature and cuts them off from many physical resources. Only when their place preferences are removed by breeding is this loss tolerable for them. Instead of being more flexible than his wild ancestor, the domestic is specialized to accept human judgment concerning habitation and food. The capacity for living with deficiency is not a true liberation of behavior but the weakening of the choice-making faculty. Wild cattle range widely over diverse types of soils and vegetation in search of plants for which they have a special need at certain times or for trace elements and other minerals that they lick directly from the earth. They seek mud or sand, shade or bright sun, humidities and winds—the conditions that are right for them. Their deliberate instinctive exposure to rain and snow is precisely regulated to their requirements. Many have special relationships with birds, who feed near or on them, and with other animal and plant parasites, internal and external, beneficial to their health. Each step in their life cycle is carried out in the right surroundings, which may be different for feeding, giving birth, courting, resting, hiding, playing, or socializing. In zoos, mental and physical breakdowns are common because the animals lack the extensive range of choices necessary to a healthy physical or social existence. Some species simply cannot be kept alive, necessitating a constant flow of "living material" to replace the dead or dying.

Domestic animals who also live in restricted environments are not stir-crazy and malnourished because they are the survivors of hundreds of generations of captives. They are the well-padded drudges, insulated by blunted minds and coarsened bodies against the uniformity of the barn-yard, having achieved independence from the demands of style by having no style, coming to terms with the grey world of captivity by arriving at the lowest common denominator of survival.

If this seems to slander some favorite dog or horse or pig, remember that artificial selection of juvenile qualities also favors immaturity, flexibility, and adaptability. The qualities that are admired—responsiveness to men in dogs and trainability in horses—are achieved through breeding at the expense of the traits of maturity. No one can judge the pathos of the domestic animal who has not watched its wild cousin in its natural habitat over a period of months. As long as civilized mythology ranks wild animals as poor relations to barnyard forms it will be almost impossible for most people to make unbiased comparisons.

Occasionally man himself is included in lists of domestic animals. But man is civilized, not domesticated. Domestication is the process by which the genetic make-up of organisms is modified by man to make breed lines and by which civilized man controls organisms that constitute part of his own habitat and through which he perceives all of nature. These lines are disengaged from the niche of the wild stock, stripped of biological integrity, simplified in behavior and requirements.

Among domestic animals social relationships are reduced to the crudest essentials. Pre-reproductive parts of the life cycle are minimized, courtship is reduced, and the animal's capacity to recognize its own species is impaired. Since these changes have not taken place in man, man is not properly a domestic animal, although civilization has disrupted his epigenetic stability and loosed a horde of "goofies."

"The Balance of Nature"

Rachel Carson

From all over the world come reports that make it clear we are in a serious predicament. At the end of a decade or more of intensive chemical control, entomologists were finding that problems they had considered solved a few years earlier had returned to plague them. And new problems had arisen as insects once present only in insignificant numbers had increased to the status of serious pests. By their very nature chemical controls are self-defeating, for they have been devised and applied without taking into account the complex biological systems against which they have been blindly hurled. The chemicals may have been pretested against a few individual species, but not against living communities.

In some quarters nowadays it is fashionable to dismiss the balance of nature as a state of affairs that prevailed in an earlier, simpler world—a state that has now been so thoroughly upset that we might as well forget it. Some find this a convenient assumption, but as a chart for a course of action it is highly dangerous. The balance of nature is not the same today as in Pleistocene times, but it is still there: a complex, precise, and highly integrated system of relationships between living things which cannot safely be ignored any more than the law of gravity can be defied with impunity by a man perched on the edge of a cliff. The balance of nature is not a *status quo;* it is fluid, ever shifting, in a constant state of adjustment. Man, too, is part of this balance. Sometimes the balance is in his favor; sometimes—and all too often through his own activities—it is shifted to his disadvantage.

Two critically important facts have been overlooked in designing the modern insect control programs. The first is that the really effective control of insects is that applied by nature, not by man. Populations are kept in check by something the ecologists call the resistance of the environment, and this has been so since the first life was created. The amount of food available, conditions of weather and climate, the presence of competing or predatory species, all are critically important. "The greatest single factor in preventing insects from overwhelming the rest of the world is the internecine warfare which they carry out among themselves," said the entomologist Robert Metcalf. Yet most of the chemicals now used kill all insects, our friend and enemies alike.

The second neglected fact is the truly explosive power of a species to reproduce once the resistance of the environment has been weakened. The fecundity of many forms of life is almost beyond our power to imagine, though now and then we have suggestive glimpses. I remember from student days the miracle that could be wrought in a jar containing a simple mixture of hay and water merely by adding to it a few drops of material from a mature culture of protozoa. Within a few days the jar would contain a whole galaxy of whirling, darting life—uncountable trillions of the slipper animalcule, *Paramecium,* each small as a dust grain, all multiplying without restraint in their temporary Eden of favorable temperatures, abundant food, absence of enemies. Or I think of shore rocks white with barnacles as far as the eye can see, or of the spectacle of passing through an immense school of jellyfish, mile after mile, with seemingly no end to the pulsing, ghostly forms scarcely more substantial than the water itself.

We see the miracle of nature's control at work when the cod move through winter seas to their spawning grounds, where each female deposits several millions of eggs. The sea does not become a solid mass of cod as it would surely do if all the progeny of all the cod were to survive. The checks that exist in nature are such that out of the millions of young produced by each pair only enough, on the average, survive to adulthood to replace the parent fish.

No one knows how many species of insects inhabit the earth because so many are yet to be identified. But more than 700,000 have already been described. This means that in terms of the number of species, 70 to 80 percent of the earth's creatures are insects. The vast majority of these insects are held in check by natural forces, without any intervention by man. If this were not so, it is doubtful that any conceivable volume of chemicals—or any other methods—could possibly keep down their populations.

The trouble is that we are seldom aware of the protection afforded by natural enemies until it fails. Most of us walk unseeing through the world, unaware alike of its beauties, its wonders, and the strange and sometimes terrible intensity of the lives that are being lived about us. So it is that the activities of the insect predators and parasites are known to few. Perhaps we may have noticed an oddly shaped insect of ferocious mien on a bush in the garden and been dimly aware that the praying mantis lives at the expense of other insects. But we see with understanding eye only if we have walked in the garden at night and here and there with a flashlight have glimpsed the mantis stealthily creeping upon her prey. Then we sense something of the drama of the hunter and the hunted. Then we begin to feel something of that relentlessly pressing force by which nature controls her own.

Everywhere, in field and hedgerow and garden and forest, the insect predators and parasites are at work. Here, above a pond, the dragonflies dart and the sun strikes fire from their wings. So their ancestors sped through swamps where huge reptiles lived. Now, as in those ancient times, the sharp-eyed dragonflies capture mosquitoes in the air, scooping them in with basket-shaped legs. In the waters below, their young, the dragonfly nymphs, or naiads, prey on the aquatic stages of mosquitoes and other insects.

Or there, almost invisible against a leaf, is the lacewing, with green gauze wings and golden eyes, shy and secretive, descendant of an ancient race that lived in Permian times. The adult lacewing feeds mostly on plant nectars and the honeydew of aphids, and in time she lays her eggs, each on the end of a long stalk which she fastens to a leaf. From these emerge her children—strange, bristled larvae called aphis lions, which live by preying on aphids, scales, or mites, which they capture and suck dry of fluid. Each may consume several hundred aphids before the ceaseless turning of the cycle of its life brings the time when it will spin a white silken cocoon in which to pass the pupal stage.

And there are many wasps, and flies as well, whose very existence depends on the destruction of the eggs or larvae of other insects through parasitism. Some of the egg parasites are exceedingly minute wasps, yet by their numbers and their great activity they hold down the abundance of many crop-destroying species.

All these small creatures are working—working in sun and rain, during the hours of darkness, even when winter's grip has damped down the fires of life to mere embers. Then this vital force is merely smoldering, awaiting the time to flare again into activity when spring awakens the insect world. Meanwhile, under the white blanket of snow, below the frost-hardened soil, in crevices in the bark of trees, and in sheltered caves, the parasites and predators have found ways to tide themselves over the season of cold.

The eggs of the mantis are secure in little cases of thin parchment attached to the branch of a shrub by the mother who lived her life span with the summer that is gone.

The *Polistes* wasp, taking shelter in a forgotten corner of some attic, carries in her body the fertilized eggs, the heritage on which the whole future of her colony depends. She, the lone survivor, will start a small paper nest in the spring, lay a few eggs in its cells, and carefully rear a small force of workers. With their help she will then enlarge the nest and develop the colony. Then the workers, foraging ceaselessly through the hot days of summer, will destroy countless caterpillars.

Thus, through the circumstances of their lives, and the nature of our own wants, all these have been our allies in keeping the balance of nature tilted in our favor. Yet we have turned our artillery against our friends. The terrible danger is that we have grossly underestimated their value in keeping at bay a dark tide of enemies that, without their help, can overrun us.

ᴁ

"Tsitika to Baram:
The Myth of Sustainability"
David Duffus

Sustainability is founded in a concept popularized from time-worn ecological theory and built upon by a treatment of the natural world that
reduces complex ecological relationships to mechanical production functions. That so few sustainable resource-use plans have been effective
stands as testimony to the differences between what we expect of nature
and what nature can provide. In the context of the use of wild living
resources, I suggest that sustainability is more myth than reality.

The ideal of sustainability arose, in the first instance, as sustained-
yield, or "scientific," management. Then, joining the drive for the better-
ment of all humankind, those ideas appeared in their modern incarnation, sustainable development.

This critique of sustainability is based on my experience on two rivers,
the Tsitika and the Baram. Although they occur in vastly different physical and social landscapes, the desire to harvest resources from these river
valleys and to invoke a sustained-yield framework transcends the differences between the sites. That leads to an uneasy observation of how deeply
we have ingrained the myth of sustainability across space, landscape, and
culture.

The Tsitika is a small river in the coastal temperate forest on the east
coast of Vancouver Island, Canada. A decade ago it coursed unmolested to
a small bay on Johnstone Strait, a deep channel between Vancouver Island
and the Broughton archipelago, islands which eventually coalesce into the
Canadian mainland. As logging progressed north on Vancouver Island, it
brought down most of the old-growth coniferous forests. In some cases,
trees were replanted, but, as in much of the coastal forest, replanting
efforts will not soon replace the forest, only the trees. The Tsitika River
meanders into the salt water at Robson Bight, a place well known for summer aggregations of killer whales (*Orcinus orca*).

Tension between logging interests and conservationists, preservationists, and native peoples grew to the point of civil disobedience and legal

action. Given the history of such events in Canada, forestry interests won the day and continued logging down the valley, only to be halted in mid-1992 by a joint committee of Canadian federal and provincial authorities who took the whales' interest to be more valuable at this location than the small stands of merchantable timber left in the lower valley.

Sustainability was invoked on two fronts in this case. Originally, industry and Forest Service planners stated that the rate of logging was sustainable. However, so many different figures and interpretations have been made of the forest's productivity in the Tsitika valley that no one group seems capable of taking an unbiased view of the facts. Now it appears that nobody believes that the forest has been logged at a sustainable rate. Yet sustained use of forests is the prime directive of the forest resource managers.

Sustainability is also a founding principle for the wildlife tourism associated with the killer whales of Robson Bight. The tourism and recreation interest will create a sustained demand. Killer whale watching has only limited substitutability. But how can one maintain a supply of whale encounters? It seems as if the whales' presence in Robson Bight is a gift. No research has provided indications as to why the whales congregate at Robson Bight, what is the character of the relationship, and what forces may alter the current pattern. These are difficult questions, granted, but the volumes of research done on the whales at this site have ignored the obvious and necessary ecological and biogeographical research, focusing instead on interesting but inapplicable descriptions of behavior and populations.

Even more demanding is to attempt to answer questions regarding potential links between land use in the river valley and whale use of the site. Is wildlife tourism sustainable? Any answer that does not address the two-fold questions of forestry and whale ecology is not addressing sustainability.

The Baram River is embroiled in a conservation issue with international bearing as selective logging alters the nature of the mixed dipterocarp forest on the island of Borneo. In this case, sustainability has ecological, social, and cultural meaning as the last of the forest people, the Penan, are brought into the web of economic development.

An international industrial coalition established a level of sustainable timber yield, but many question its ecological relevance. The dynamics of tree replacement in selectively logged tropical forest is a complicated issue. When patches open up, new trees grow, but will they produce a future for forestry or reproduce biological communities that fulfill any other user's demands? Some species will flourish in newly opened patches; some will disappear. Given the propensity for speciation in these fine-grained ecosystems and the resulting diversity, forestry practices may risk hastening extirpations, if not extinctions.

Furthermore, the sustainability of the social system is highly suspect. People from the heterogeneous cultural groups in the Baram drainage have moved into the labor economy, and many groups are quite pleased with the economic opportunity. Once the boom is over, will the cultural system, including the economic, political, and social systems, be able to deal with the relaxation accompanying a bust? More important, will the forest be able once more to provide the types of sustenance and trade goods it had before logging? And, perhaps the most thorny issue, will the indigenous people who use the forest most directly, the Penan, be adequately informed and compensated for those losses?

Forces on both sides, the Malaysian government and environmental groups, contend that they are dealing sensibly with the forestry issue and the Penan issue. The situation is so polarized that both are likely missing many subtleties, such as the sensitive issue of rate of change in indigenous lifestyles, and the relationship between people that may be strained by the secular treatment of one group. Clearly, many decisions have been made regarding the Baram and its indigenous people in the context of great ecological and social uncertainty. If the social consequences of economic activity are not addressed, then sustainability is not addressed.

The ties that bind the Tsitika River in temperate, industrial Canada, and the Baram River in the tropical, developing landscapes of Sarawak begin at a fundamental level. Human interaction with nature in both cases is firmly rooted in an ethic of sustainable use. Both river basins have experienced rapid environmental change that has complex social and ecological consequences; coincidentally, this change involves forestry. In both cases, large corporations play pivotal roles in the rate and style of resource exploitation, and there are links between the corporations and state governments that tend to enhance current business profits.

More troubling is the bipolar response of the public. In both places the public, at least those who show concern, is divided into two camps. The people who are in a position to reap immediate and direct benefit from forestry form support groups to promote the government–corporate position. Facing off against them is a coalition of environmentalists and indigenous people. Both sides cry foul and use the media to harp self-righteously over the injustice. Generally, the environmental groups gain more membership and thus financial support from the publicity, and the forest industry continues logging.

Meanwhile the Tsitika rumbles through the past winter's mudslides and logging debris from new blowdowns, more frequently spewing silt into the tidal current of Johnstone Strait. The Baram rises and falls more vehemently, carrying tons of suspended sediment in its flow. Log booms and trucks continue to carry trees out of both valleys as if they will never run out. What and who carries the burden from all this change? First and foremost, the cost is borne by the inhabitants of each valley. Whether they be wild pig or timber wolf, hornbill or spotted owl, some as yet unnamed species of forest ant, or a local wildflower squeezed by the forces of evolution and isolation into a new, precarious species, they all bear the cost. When losses accumulate to the population level, species are lost.

As a society, we have difficulty putting that loss of species into perspective. Each species is an expression of fundamental life forces, yet it is also dynamic. All species are headed for extinction in the long run, but if replaced by the same forces of nature, any single type of plant or animal is not in itself significant but the player of a significant role in its own time and space. Any prescription that does not acknowledge the context of evolutionary force does not address long-term sustainability.

Sustained production for human use has undeniable roots in evolution and ecology; economics and social factors are relevant only after the fact. Yet the reams of material now being passed off as sustainable-yield management, sustainable development, and sound resource-use planning do not but pay lip service to either ecology or evolution. Unfortunately, although the public, politicians, and bureaucrats shudder at the idea, we are critically lacking in fundamental knowledge regarding living resources. We need far more research; we need research done by scientists thinking like philosophers, and philosophers thinking like scientists—in other words, the work has to extend beyond the traditions of secularized knowledge as it has developed.

Trees and rivers are parts of perfect circuits of nutrients, water, and genes. Standing at the mouth of the Tsitika River watching the warm river water and the frigid ocean water mix, or flying over the sultry coastal plain where the Baram pushes over its bank into impenetrable neepah palm swamps, I do not feel impatient about scholars and scientists finding answers; rather, I feel admiration for those who have even undertaken the task of answering questions about the workings of the natural world.

For places like the Tsitika and Baram Rivers, the message is clear. The theme behind the utilitarian ethic that has driven our quest for more production from the forests, grasslands, and seas—that the fruits of the earth are always there for us to use—is dead wrong. And it is not simply a matter of adjusting the rates at which we harvest things. Evidence is mounting that there are some forests we cannot log, some fish we cannot catch, some streams we cannot pour effluent into, some places where humans can tread only lightly. The Tsitika and the Baram are places where the scientifically threadbare nature of the myth of sustainability plainly shows through. Although this will find little favor given our species' seemingly inherent arrogance, humility is perhaps the most important survival tool we have left. One wonders where, then, is the well-spring of humility?

"Toward a Science of Letting Things Be"
Bill Willers

The following written exchange between Bill Willers and Daniel T. Blumstein appeared in the journal Conservation Biology *in December 1992, June 1993, and September 1993.*

In a piece written in 1990, David Brower, with regard to the few remaining remnants of pristine landscape, noted that "there are still some fair ladies, but too many faint hearts to succeed in winning them." Indeed. The problem he cites extends deeply into the ranks of conservation biology, where too many hearts are not only faint but also anthropocentric, managerial, and absolutely brimming with the moderation that has brought us to our present environmental condition.

After reading a special section on the Greater Yellowstone Ecosystem (*Conservation Biology* 5:3), I came away deeply concerned about viewpoints expressed in a number of the articles. There were calls for greater compromise from conservationists, for a determination of the Ecosystem's fate by local residents, and for more data and greater specialization in the interests of proper management. There was also suggestion that wilderness advocates are inclined toward acting out fantasies related to nature. Much of what was written seemed to me to derive from a lack of appreciation of just what concerns surrounding conservation biology are all about.

Biodiversity refers not merely to a wide range of genetic information but also to legions of interspecific relationships, the effects of which reverberate within dense meshworks of process not fully understood. In robbing an ecosystem of its top-level predators and in then thinking up strategies for handling overpopulation of prey species; in permitting timbering, mining, and the grazing of domestic stock; in failing to combat roading and fragmentation, humanity molds an entity much as it might train a dog. And this opposes any concept of a reserve in which natural process proceeds unimpeded.

To dwell endlessly on the tasks of obtaining more and ever more data for the expressed purpose of managing a biological reserve is to suggest that enough knowledge for such a task is just around the corner. That is not so. Much of the detail regarding processes operating in large ecosystems eludes human understanding still—and probably will into the foreseeable future. The point, though, is that if left alone so that its processes can continue in an unmanaged way, a vast ecosystem, unlike a trained dog, becomes a teacher.

When biologists concern themselves with the fate of public lands they come up against bureaucrats, politicians, and profit-driven resource users, all of whom together compose a power bloc with tendrils extending through Washington's Beltway and into corporate boardrooms. It is a bloc not remotely concerned with natural process unless that process stands in the way of its agenda, which is obvious enough that it needn't be elaborated on here. It is important, though, to point out that this power bloc lives in a world of confrontational politics, of rough-and-tumble business competition, and of information manipulation—a world in which academic biologists are neither comfortable nor skilled. When biologists deal with such a power bloc in a spirit of compromise there is no outcome that could ever result in a world biological reserve worthy of the title.

Anyone truly interested in preserving diversity must cling to principle and learn to play hardball. Regarding the Greater Yellowstone Ecosystem, the vast bulk of which is presently national forest, a good first step toward status as a world reserve would be to send the Forest Service and its multiple-use philosophy packing, never to be heard from again in the region. The name or uniform of what replaces it isn't important. What is important is that ultimate guidance come solely from a committee of broadly-educated biologists who are free of political and industrial domination, and who are firmly committed to a return to wilderness conditions and to the protection of organic evolution.

The full range of process appropriate to large ecosystems will either be allowed to survive on the planet or it will not. There is no middle ground here. If that which has functioned beautifully through the eons free of human meddling is to survive, "management" must become an erasing, a reversing, a minimizing of human impact—a science of letting things be.

If biologists won't fight hard for pure process, nobody will. And if as a group they are going to be content with a grazed, mined, clearcut, roaded complex, they should begin looking for a designation that fits—such as "arboretum" or "zoo." To refer to the Greater Yellowstone Ecosystem in its present form as a world biological reserve is simply intellectually dishonest.

At this time the biological community is well armed with objectively-acquired information. It should be able now to stand against the political–industrial juggernaut that has always called the shots on public lands, and to advocate from a position of strength on behalf of the natural world. Moreover, it should be concentrating on the task of moving all of these issues relating to evolutionary process out of their in-groups and into the public arena.

Doris K. Miller wrote in 1972 that "by taking no position [scientific] societies, in effect, take a very influential position." One could well expand on that view and say that in taking a weak position scientists are being influential. For when scientists who allegedly stand for conservation take a weak and compromising position they become not only allies of destructive forces but also drogues acting against a healing tide that is trying desperately to gain momentum.

Do this: Take a powerful, rock-steady stand for pure process. That failing, stand aside and refrain from neutralizing those who will.

"Making Humans Part of the Solution: A Reply to Willers"

Daniel T. Blumstein

With some trepidation, I feel compelled to respond to Willers's (*Conservation Biology* 6:605–607) call for the removal of humans and human impact from wilderness and natural systems. My trepidation comes from my belief that the best policy is ultimately formulated as a result of having extremely diverse viewpoints represented. Willers has taken an extreme viewpoint on the role of humans in natural systems. For the most part I agree with his assertion that we can learn a lot about nature by observing natural systems, and I think I share what I perceive to be his deep love and respect for nature. I reply because I feel his attack on those who seek to manage natural resources pragmatically is overgeneralized, potentially impractical, and potentially imperialistic. My response will deal with two of his implicit assumptions: that by removing human impact we preserve natural systems and that it is possible to manage a natural resource without the cooperation of local peoples.

First, Willers implies that human impacts make natural systems less natural, and by definition, human impacts degrade a relatively pristine location. While humans have existed as a species for an infinitesimally short period of time, we have undeniably done our share to influence the ecology of Earth. However, I find fault with Willers for his overgeneralized assertion that conservation policy should *always* try to remove human impact. While I would agree that excessively destructive uses should be eliminated or minimized in pristine areas, sustainable uses may not only maintain diversity but may be crucial for the immediate survival of focal species. An example may illustrate this point.

I've been working in Khunjerab National Park, a large, high alpine national park located in Northeastern Pakistan. Khunjerab has no permanent inhabitants and is said to have remarkably dense populations of snow leopards. Snow leopards are said to rely heavily on the livestock grazed in the park during the summer. In the fall, shepherds drive their herds out of the park to winter in lower-elevation villages. Some snow leopards descend to the villages to eat the livestock and, if caught, are killed by the residents. Few snow leopards are said to be killed in the park. I suggest that snow leopard densities may have increased from a long history of livestock grazing in this area. Clearly, livestock grazing has a myriad of impacts on a natural system. However, in this case grazing may be consistent with a policy designed to sustain snow leopards. Snow leopards are found scattered throughout Central Asia. The loss of each additional population probably reduces the chances of the long-term survival of the species. In general, I feel that Willers's assertion that relatively pristine areas should have no human impact needs critical questioning prior to recommendation or implementation.

Second, Willers was concerned about the future of the Greater Yellowstone Ecosystem because conservationists (see *Conservation Biology* 5:3) called for "a determination of the Ecosystem's fate by local residents" (Willers 1992: 605). Whether we like it or not, people living in an area exert great influence on the ability of conservation biologists to manage a natural resource in that area. Residents can and do fight unappealing legislation through legal and sometimes illegal means. Laws may be challenged in court, or simply violated following enactment. While I personally may take a more authoritarian role when dealing with a conservation problem in a developed country like the United States (particularly when "big business" is involved), I feel that on an international level, specifically when dealing with less developed countries, successful conservation biology cannot be divorced from development, lest we, as well-meaning conservationists, become conservation imperialists.

Development may be defined and implemented in many ways, but the essence of its effect is to raise the standard of living for people living in relatively impoverished conditions. It is too easy for those living in posh conditions to dismiss the needs, wants, and desires of the majority of people on Earth. The wise management of natural resources may be a viable and sustainable route for development. Many less-developed communities rely on natural resources for their very existence, and may only be destroying them for a quick influx of cash.

Participatory development, a current buzzword in international development, is designed to involve local communities in their own development. If natural resources are particularly unique, then a local ecotourism industry may be developed, and both residents and global citizens may share the resource. Less unique resources may be managed in a way that helps both the residents and protects the "organic evolution" that Willers advocates.

I feel that humans should be viewed as part of the solution and not as part of the problem. Wise management of natural resources should assess and address both the role of humans on resources and the desires of those living closest to resources.

🐾

"Ecocentrism versus Management: A Reply to Blumstein"

Bill Willers

The following is a reply to Daniel Blumstein's letter (Conservation Biology 7:223–224) about my earlier piece on diversity (Conservation Biology 6:605–606).

My view that some areas of the planet should be kept free of human "management" is not, as was suggested, imperialistic. The assertion that it is, potentially, is incongruous given some of Blumstein's comments (such as, "Whether we like it or not, people living in an area exert great influence . . . can and do fight unappealing legislation through . . . sometimes illegal means. Laws may be . . . simply violated," and ". . . conservation biology cannot be divorced from development . . .").

I did not mean merely to imply that human impact makes natural systems less natural. I thought I had succeeded in stating that flatly. At this late date I refuse to accept the image of humanity as hunter–gatherer or simple goat herder. Though such remain in areas, humanity's direction is clearly toward a technoworld, even as pundits are predicting a doubling of the world's human population in a couple of generations.

Regarding the Yellowstone Ecosystem, residents there are anything but impoverished. They are ranchers, miners, lumbermen, and tourist merchants who live very, very well, and who, with their development ethic, their mushrooming towns, and their unceasing push for economic growth, are killing what is allegedly our premier wilderness.

The "jobs" argument, where it relates to public lands, is bogus. Title to such lands is in the hands of 250 million citizens who have no moral obligation to maintain the jobs and lifestyles of a minority whose work is destructive to the land base. Do we struggle to keep drug pushers in business just because their work is gainful and pays the rent? When "jobs" no longer serve society, retraining is in order.

Blumstein's comment that "wise management of natural resources may be a viable and sustainable route for development" sounds like a Ron Arnold quotation right out of the Wise Use Movement. What is being said here, since all I have argued is that *some* areas be allowed to function without human management, is that *all* areas should be managed. I disagree wholeheartedly. Nor do I agree that it is "an extreme viewpoint," as Blumstein states, that some corners of the planet should be allowed to function without human intervention. Indeed, it is a very conservative viewpoint.

Blumstein's example ("Clearly, livestock grazing has a myriad of impacts on a natural system. However, in this case grazing may be consistent with a policy designed to sustain snow leopards") is not at all persuasive. In fact, it supports my view, because it looks at things in isolation. In this scenario, one facet of an ecosystem, a cat, is managed for, with the result that everything else that happens—all ripple effects within the community—gets lumped into the rather broad category of "myriad of impacts on a natural system." And that focuses nicely on my fundamental question: When we have succeeded in the "wise management" of everything we can get our hands on, how, in the name of good sense, will we ever be able to know what a natural system is?

I have read Blumstein's arguments, and I am unmoved. To manage *everything,* I maintain, is not "wise."

From
"Down from the Pedestal—
A New Role for Experts"
David Ehrenfeld

We live in a society that worships expert knowledge, and sometimes worships the experts themselves. This can be very pleasant if you're an expert, but has several curious and important drawbacks for the rest of us.

The first and maybe the most important drawback is that expert knowledge is rarely sufficient for analysis, prediction, and management of a given situation, with the exception of basic repair operations. This is because in order to limit the number of variables they have to contend with, experts make the assumption that the systems they are working with are self-contained, defined, closed. Real-life systems are hardly ever closed. Naturally, some systems conform better to the assumptions of the experts than others; they are closer to being closed. So there is a range of success which is as much dependent on the field as it is on the expert him- or herself. At the dismal end of the range are those open-system fields where expert knowledge, especially concerning long-range prediction and design, is worth less than its face value, and is sometimes not much more useful in making sound decisions than a random number table. In these fields, expertise can be a pose, heavily dependent on professional jargon and a smokescreen of mathematics, statistics, and technology. The pose usually works. Some obvious examples are economics, long-term weather forecasting, and educational psychology.

At the other end of the range, where the expert's assumptions of a closed system can approximate reality, are such fields as routine dentistry and air traffic control. Some of us might not want to be dentists or air traffic controllers, but we have to admit that, given a reasonable degree of professional competence, they usually get the job done as specified in advance.

Where do the expert fields of ecology and wildlife management fit in this range? Certainly they are not as bad as economics or weather forecasting; there is more ability to make pretty good, long-term, or at least mid-term, predictions that show trends in the right direction and within an order of magnitude of the actual numbers. On the other hand, the science and management of natural resources are a far cry from dentistry or television repair in the correspondence between predictions and actual events or in the fulfillment of planned objectives. In his celebrated, 1976 keynote address to the annual meeting of the American Fisheries Society, in Michigan—an address entitled "An Epitaph for the Concept of Maximum Sustained Yield" and directed towards an audience of expert fisheries biologists—Philip A. Larkin began to define the problem:

> Natural systems are sufficiently diverse and complex that there is no single, simple recipe for harvesting that can be applied

universally. When there is added in the complexity and variety of social, economic, and political systems, the number of potential recipes is just too enormous to be easily summarized by simple dogma. Perhaps the best we can hope for is a general statement of principles with accompanying guidelines that should be applied in the hope of ensuring that we will trend in the best direction.

Although Larkin's paper was about a specific management issue, I think that the quotation has general usefulness. The reductionist methods of science, which can work extremely well in closed systems, tend to break down under the open-endedness imposed by biological complexity and by the interacting complexities of political, economic, and social factors.

⚮

From
"Some Thoughts on Ecological Planning"
Raymond F. Dasmann

In what I have come to see as an ecologically "ideal world," human settlements form an archipelago of urban or agricultural islands, connected by transportation and communication routes, but set within a matrix of wild country. By wild country I mean landscapes in which natural vegetation and animal life can continue to thrive without dependence on human assistance or interference. This surrounding wild matrix could be used in part to provide materials for human use, from fodder for livestock, lumber for housing, or minerals for industrial use, so long as these uses were not so intensive as to interfere with the wild character of the country-side—meaning its ability to maintain and reproduce its natural character. In some parts of the world this pattern of human settlement persists today. Not very long ago it was the predominant pattern. Its sustainability has been tested through centuries or millennia of human use. It is probable that it could continue indefinitely if not destroyed by outside human pressures.

Regrettably, in many parts of the world, because of increases in human population and technological advances, the opposite pattern prevails. Such wild country as still exists consists of islands set within a matrix of urbanized or intensively used agricultural land, usually without connecting corridors over which wild species can move.

🦌

"A Question and Two Answers"

Lucy Redfeather

The following question, for which there is no "right" answer, was given to university students in a course on environmental issues.

Given: A forested mountain range that has never been managed or inhabited by humans. In it there has been only through-passage on foot. Indigenous people have considered it to be sacred and have left it alone so that it could function according to its own internal mechanism. The autonomous nature of this wilderness has conferred upon it a mystique so profound that even some modern technocrats sense it.

In time, the New World Order's demands for raw material and "sustainable development" triumph, and the wilderness gets "resourced." A network of roads is constructed so that the logging, mining, and grazing industries can have their way. Hunters, poachers, and operators of off-road vehicles easily access the interior, and, in time, they cause the disappearance of several species of predatory animals.

Assume that after a long spate of being "resourced," this former wilderness mountain range is vacated. Say that society demands that it be made a legal wilderness so that it can return to its former level of autonomy. Now, what would it take—and how long would it take—for this mountain range to regain the "sense of mystery" it possessed before it was entered and made available to extractive industry? What are the criteria for your opinion?

> *Here follow, somewhat edited, two answers by students. One stresses biology, while the other approaches the question from the standpoint of human perception.*

BECAUSE MANAGERIAL ACTIVITIES have changed internal biological dynamics to such a degree, it would be necessary to allow the ecosystem to return to a level of complexity and autonomy that originally existed. Species made locally extinct, if unable to recolonize naturally by migration from other populations, would have to be reintroduced. The various successional stages characteristic of that particular bioregion would have to be allowed to run their respective courses and to give way, ultimately, to climax conditions. Moreover, there would have to be allowance not only for the life span of the longest-lived climax trees but also for the entire process of their decay, since these members of the forest community provide habitat, substrate, and nutrient matter at all stages, including their condition as rotting woody debris. Therefore, in an ecosystem with long-lived trees (or any other life forms that may have been affected), one might have to think in terms of millennia.

"MYSTIQUE" DESCRIBES HUMAN PERCEPTION. Therefore, irrespective of mechanical (i.e., biological) considerations, enough time would have to pass so that society could place the period of managerial experience into a distant past. At a minimum, all members of the society who had been alive during the managerial episode would have to have passed from the scene. More probably, though, enough time would have to pass that no living member of the society would be able to recall, even as a youth, an elder who had lived during the episode. In this way, the period of "resourcing" could pass into a historical time frame so distant as to earn some mythic quality. Only then could a true mystique of the sort one associates with wilderness—with self-regulating nature— return.

❧

"What Do We Really Want From the Woods?"

Robert Kimber

What moves us is love for our home turf, love for a river, love for a cluster of mountains, for steep, wooded valleys, for the headwater brooks rising there. It's respect and affection for places, for the shape and lay of the land and for the plants and animals that live there, for the sounds and smells of those places and the feel of them underfoot, the granite, the duff of a pine stand, the cushiony mosses of a cedar swamp. We are "environmentalists" not because we love some abstraction called "the environment" but because we love places in the world the way we love our mates and our children. We love them in the flesh and because they are flesh: the creation, palpable, present, alive.

But we have learned that the existence and pleading of these feelings counts for little or nothing in a world ruled by money. We have learned that when environmental victories are won, they are usually won on economic grounds, on legal technicalities, or by good luck. So whenever we old and somewhat jaded hands go to legislative hearings on forestry reform bills and to hearings of the Land Use Regulation Commission, we too play by the established rules. We talk about economic sustainability; we stress that the appearance of our forest lands is as important to the tourism industry as sustained yield is to the wood-products industry. We talk about the economic importance of wildlife habitat and flourishing fisheries.

If we step outside these narrowly drawn boundaries, it becomes instantly clear that almost all of us have a much richer relationship to the forest than the rules of the debate let us express. But if we belong to the forest-products camp, we can't publicly talk a lot of maudlin nonsense about rocky rills and bunny rabbits, nor can we break ranks in the face of attacking green hordes. And if we are members of those green hordes, we can't talk maudlin nonsense either, because money and science are where it's at.

So it is not in the public arena but in one-on-one talks over beer or coffee that the players say what is really on their minds. A forester says he feels ashamed if he can't look at his cuts and say the woods are still beautiful. A man whose employment should place him among the vicious ravishers of the land speaks of setting aside wild land as a means of "saving Maine's soul." Another industrial land manager tells me he spends some of his happiest moments just sitting out in the woods.

These expressions of value, emotion, and delight do crop up occasionally in even the most sober and hard-headed gatherings, but they register as little more than road bumps. They lurk on the periphery of the discussion. The communal response is essentially, "Well, yes, that's all very nice, but now let's get back to business." And business is the reductionist task of asking science to tell us how little land we need to set aside to preserve existing native species and communities, and what minimum restrictions we need to impose on industrial forestry to prevent its self-destruction.

For some people, the term "biological need" may well carry with it the ecologist's deep conviction that the natural world and all its denizens—from microbes to rocks to giraffes—all have as much intrinsic right to exist as we do. But in conflicts between the rights and needs of rocks and the perceived economic needs of human beings, the rocks will always lose because money talks and rocks don't. Biology—or nature—does not have "needs." If we humans manage to reduce the earth to a rubble pile inhabited only by cockroaches, that impoverishment will be nature's response to our folly. We will lose, but "nature" won't.

So when we talk about "biological needs" we do better to recognize them as our own needs, to accept the utterly self-evident proposition that the natural world matters to us. Utilitarian value usually kicks off the list—lumber, paper, chopsticks, pharmaceuticals, the indispensable disposable diapers. But my sense is that what matters far more than any wonder drug science may yet discover in the jungles of Borneo are those aesthetic and spiritual values we choose to exclude from public debate.

What is intriguing is the growing testimony of scientists on the importance of these values. What many scientists have observed in themselves E.O. Wilson has called *biophilia*, "the innately emotional affiliation of human beings to other living organisms. Innate means hereditary and hence part of ultimate human nature." It was Wilson's personal experience that led him to offer this hypothesis, the same experience we all have when we respond to the natural world. When E.O. Wilson writes: "Life around us exceeds in complexity and beauty anything else humanity is ever likely to encounter," he confirms, on the basis of far greater understanding than most of us have, what the rest of us intuitively know: The ongoing incremental loss of wild land and the richness of life it contains brings with it an impoverishment of the spirit.

Taken by themselves, emotion and intuition are unreliable guides, but public policy that ignores what I would call the sober, well-founded passion—the dispassionate passion—of many first-rate scientific minds is hardly rational either. Feelings may not be "rational," but they tell us things it is irrational to ignore. In deciding the fate of Maine's forests, or any forests for that matter, they tell us that if we bend entirely to the demands of a global economy that insists on making and selling an excess of products we could all happily do without, we will be cutting ourselves off from our deepest roots.

"We are a wild species," Wallace Stegner wrote. "One means of sanity is to retain a hold on the natural world, to remain, insofar as we can, good animals." We should heed both the poets and the scientists because their testimony bears not only on what kind of forests we want to have but also on what kind of people we want to be. In wildness is the preservation of our own sanity, of our capacity to be good animals, or our ability to thrive as individuals and as a species.

What many voices from diverse quarters are suggesting is that we need big reserves if we want to preserve our wild, biotic heritage, which is an inextricable part of our human heritage. But no matter what science tells us and no matter how compelling the information it provides, science cannot make our decisions for us. What we decide to do ultimately belongs in the realm of ethics. Our tug of war over economic impacts and biological need fails to answer questions that our ethical, philosophical, poetic, and religious minds keep asking. Ironically, the very impulses we relegate to the periphery may be as much a part of our biology as our blood and bones.

Homo sapiens is clever in many ways but an awfully slow study in others. What is finally beginning to sink in, however, is that treating the biosphere purely as an economic resource diminishes and impoverishes us as human beings. It is the wild world that fires the human imagination, both scientific and artistic. So the questions we need to ask are not just how much land do we need to feed lumber mills and to preserve representative biotic communities, but how much do we need to leave alone if we want to be good and humble brothers and sisters to our fellow animals, if we want to keep imagination alive, if we want to remain fully human.

Part Two

Wildness and Human Society

"Repression of the ecological subconscious is the deepest root of madness in industrial society."

—Theodore Roszak

Awareness of society's connections to the natural world—connections too long ignored or forgotten—is being reawakened as wild places disappear on all fronts. Social and political movements around the world increasingly declare their identification with the natural world and concern for its integrity. Their concern has serious practical aspects. "When wilderness has been consumed," writes Wayland Drew, "our understanding of what is natural can be changed as required, and no facet of the human psyche and biology will be left invulnerable to revision."

As the wilderness shrinks away, its rarity is beginning to increase its preciousness in the public mind, and this naturally draws increased attention. This attention, and the renewed appreciation that it generates, will be the driving force behind efforts at wilderness restoration.

"Killing Wilderness"

Wayland Drew

"Oh, how great and divinely limiting is the wisdom of walls. This Green Wall is, I think, the greatest invention ever conceived. Man ceased to be a wild animal the day he built the first wall; man ceased to be a wild man only on the day when the Green Wall was completed, when, by this wall we isolated our machine-like, perfect world from the irrational, ugly world of trees, birds, and beasts."

Written in 1920, Eugene Zamiatin's novel *We,* quoted above, has never been published in its author's homeland, for the Soviet authorities quite correctly saw it to be subversive and dangerous. It describes a perfect, man-made environment, a cool, regimented, self-regulating utopia where the citizens, or Numbers, are entirely happy. Passion, ecstasy, rage, agony, heroism, and honor, all extremes by which humanity once acknowledged and enlarged its animal inheritance have been systematically reduced to a ubiquitous Good. For happiness, Zamiatin's citizens have cheerfully traded their freedom. They are secure in the knowledge that the State will meet their every need, because the State will eliminate needs it cannot fulfill.

We is the first of three great anti-utopian novels to appear in English in the last half century. Both Huxley's *Brave New World* and Orwell's *Nineteen Eighty-Four* are indebted to it, although all three books share a libertarian tradition that reaches back beyond Rousseau and the Romantic poets, a tradition exulting man's natural heritage in the face of encroaching Mechanism. Specifically, what these novels say is that a technological society will be totalitarian regardless of what political structures permit its development, for the essence of technique is efficiency, and the autonomous individual, apt to be skeptical, irrational, and recalcitrant, is inefficient. For the general good, therefore, the dangerous elements of individuality must be suppressed, and man must be severed from all the spiritual, intellectual, and emotional influences which might promote dissent. Man's integrity must be broken. He must be fragmented and reshaped to participate contentedly in the smooth functioning of the technological State—a State that is fundamentally inimical to his instinct and insulting to his intellect. In other words, the nature of man must be changed.

The protagonists of all three novels undergo this change and although the techniques vary they are uniformly relentless. The issue is never in doubt. "Reason," says Zamiatin's hero as he awaits his lobotomy, "must prevail." Since these are visions of perfectly rational States, it is clear that for the novelist freedom consists largely in irrationality, in instinctual response, and in the right to reject oppressive but reasonable options. Some people in *We* have retained the right. They are those who live in the wilderness beyond the Green Wall. The inhabitants of the State who know of their existence fear them deeply, for they pose a radical, primitive, viable alternative to the ethos of uniformity. In fact, the wilderness itself offers such an alternative. Vast and turbulent, it constantly invades the sterile, constructed world with reminders of its presence . . . "from some unknown plains the wind brings to us the yellowed honeyed pollen of flowers. One's lips are dry from this sweet dust. [It] somewhat disturbs my logical thinking." In its mystery and diversity, in its exuberance, decay, and fecundity, the perfection of the wilderness contrasts with the sterile and static perfection of the State. The difference between them is that between existence and life, between predictability and chance, between posturing and action. Wilderness, Zamiatin says, will threaten the totalitarian state while they co-exist, for the separation of man from nature is imperfect so long as man might recognize that a separation has occurred.

Zamiatin knew a good deal about the conquest of nature, for he was a civilized man. But to a Russian writing fifty years ago, the utter technological crushing of the wild and the free was inconceivable. He therefore assumed that State control would advance mainly on one front towards the subjugation, fracturing, and reconditioning of the individual. Huxley made the same assumption, but it is interesting to note that in *Brave New World* wilderness has been drastically diminished, to the point where Green Walls are no longer necessary. At the same time human techniques have been refined to near-perfect efficiency. "A love of nature," says the Director of Hatchery and Conditioning, "keeps no factories busy. . . . We condition the masses to hate the country . . . but simultaneously we condition them to love all country sports. At the same time, we see to it that all country sports shall entail the use of elaborate apparatus. So that they consume manufactured articles as well as transport."

In George Orwell's dreadful vision, written sixteen years later, man has been unraveled from the fabric of nature. Parks remain, where citizens might take collective hikes under surveillance, and a few pockets of wild land still offer seclusion. In one of these forgotten corners, reminiscent of the Golden Country of his dreams, Winston Smith first makes illicit love to Julia. "It was," Orwell tells us, "a political act," because it was instinctual and therefore subversive. Elsewhere, only memories remain, and truncated passions, and hopeless atavisms, all of which can be easily excised or altered by human techniques. "If you want a picture of the future," says O'Brien, the Thought Policeman, "imagine a boot stamping a human face—forever." In the context of *Nineteen Eighty-Four,* he is absolutely right; there is no escape.

Monitory novels should be read in groups, one after the other, for then their various crosscurrents are less diverting and the reader is better able to sense the drift of his own society. Many such novels have appeared in recent years, but these three remain predominant (*Brave New World* and *Nineteen Eighty-Four* together still head their publisher's list in Canadian sales). Huxley's vision of the future, forty years old this year, is closest to the present truth, for we have in fact passed beyond the necessity for Zamiatin's Green Wall and we have not yet realized the Orwellian nightmare. We are at the stage where, to quote one of *Brave New World's* Controllers, "People are happy. They get what they want and they never want what they can't get. . . . [T]hey're so conditioned that they practically can't help behaving as they ought to behave." As for wilderness, it is seen as an archaic, anarchistic welter. When its mystique has been evaporated, its measurable components, such as water, oxygen, minerals, timber, space, lie open to the service of technocracy.

That technocracy operates, as Huxley predicted, with subtlety and refinement. Its workings have been carefully traced by Jacques Ellul[1] and Herbert Marcuse.[2] Its dynamic is directed toward no less an end than the sterilization of the natural world and the substitution of total predictability. When it is understood that we are in its grip, the remaining wilderness assumes an awesome importance, for it is the sole index by which we can measure the extent of our own subjugation to unnatural forces. When wilderness has been consumed, our understanding of what is natural can be changed as required, and no facet of the human psyche or biology will be left invulnerable to revision. Reason, and only Reason, will prevail.

The South African novelist Laurens van der Post recently posed the challenge succinctly: "It is not reason that needs to be abolished," he said, "but the tyranny of reason."[3] But for the contemporary, existential urban man constantly assaulted by novelties, diversions, and facile, conflicting opinions, such a statement is already meaningless. What is reason if not consensus? And how can any tyranny exist in such a proliferation of choices, such an unprecedented prosperity and scope for self-expression? Already for millions of such men the rationale of the technocracy has become absolute, and the highest use of intelligence consists in maintaining their position in it. To be sure, their lives are fraught with problems and dilemmas, but none of these is insoluble within the terms of the artificial environment, an environment sufficiently elastic to desublimate repressed instincts in harmless ways. Promiscuity, drugs and alcohol, gambling, movies and television, violence and combativeness in sports and games, all are thus enlisted in the State's service. They divert and purge simultaneously, as do the debates generated by their presence, thereby obscuring criticisms of technocracy itself. Meanwhile, the Reason of the technocracy grows stronger by self-confirmation, for it can easily be shown that technological problems demand technological solutions. Everywhere we are acceding to the technocratic dictum that what is not known by experts cannot be known.

Only in wilderness is it possible to escape this tyranny. In wilderness a man or woman has physically left behind the milieu of conditioning—the pervasive sociability, the endless "information" from mass media, and so on. To some extent, the wilderness traveler will be reminded of his animal nature, and share again the profound, irrational correctness of trees, lakes, birds, and beasts. For urban men this can be a subverting experience. Some must react violently in an attempt to debase or destroy the source of their disturbance, and to bring ancient terrors to heel. But even on the most superficial level wilderness strengthens independence, for the man who has been freed from regimentation and finds that he can go anywhere at any time has been reminded of a basic animal right. Should he succeed in formulating the idea of *right,* then in a small but significant way he will become a critic of technological confinement. There is a fundamental difference between this animal freedom and technocracy's most popular accomplishment: the ability to travel thousands of miles in a regulated atmosphere, never once feeling the rain or the sun, never once drinking pure water, hearing a natural sound, or breathing unreconditioned air. The wilderness traveler is apt to find himself in a radical position, for he has passed beyond the "reasonable" arguments about public versus private transportation, or jumbo jets versus the SST, or whether or not we are economically capable of mass producing a safe automobile. He has bypassed the mass of alternatives posed by the assumptions of the technological society and glimpsed a possibility, which his society will tell him is reactionary, archaic, and impossible, but which his body and his spirit tell him is absolutely correct. He has positioned himself to breach the Reason of his society, to jump the Green Wall and confirm that there is something better then being a drugged and gratified utopian.

> The man of flesh and bone can maintain physical and mental sanity only to the extent to which he can have direct contact with a certain kind of reality not very different from the conditions under which he evolved.[4]

As the anti-utopian novelists foresaw, a force bent on total control must first confuse the inherited biological indices which tell us what types of behavior and what forms of environment are consistent with the dignity and survival of the human animal. The conservationists who now oppose that force recognize that the proper exercise of reason includes the defense of the instinctual and irrational, both inside man and in what remains of the natural world. Such people see in the issue of wilderness preservation a chance to negate what subjugates and diminishes them as individuals. They are saying in effect that they prefer freedom to happiness, even now. Like the Savage in *Brave New World*, they reject surrogates, and defiantly claim the right to God, to poetry, to real danger, to freedom, to goodness, to sin.

> "In fact," said Mustapha Mond, "You're claiming the right to be unhappy . . . not to mention the right to grow old and ugly and impotent; the right to have syphilis and cancer; the right to live in constant apprehension of what may happen tomorrow; the right to catch typhoid; the right to be tortured by unspeakable pains of every kind."
> There was a long silence.
> "I claim them all," said the Savage at last.

But it is one thing to have attained such a perception and quite another to know how to act upon it. Flight is still possible for us, as it is for Huxley's citizens, but most have been conditioned away from the necessary decisiveness and courage. Besides, at the present rate of technological expansion, escape could only be relative and temporary.

Environmental defense within the society seems to offer the larger hope. Traditionally, conservation has selected goals not incompatible with the objectives of the society at large—a stretch of marshland, a grove, a sand-spit, a strategic watershed, a particular species of endangered bird—such concerns coincide with the fragmenting process of technology and do not seriously threaten its advance. In fact, the stronger conservation has become the more it has hastened refinement of human and management techniques relative to land use, and the recent enlargement of its vision to include the Earth itself tends merely to reinforce the apparent need for tighter, global, technocratic controls. The threat of breakdowns in ecological systems can only be countered "realistically" by urging either the totalitarian management or replacement of those systems. "Spaceship Earth," a current catch phrase among environmentalists, indicates their co-optation by the technological rationale, for the spaceship is the absolute in technical perfection. In its operation there is no room for the irrational and nothing can be left to chance. The survival of those who inhabit it depends on their subservience to technical processes, and hence on their diminishment as humans.

What conservation activities have accomplished, however, is the stubborn keeping alive of a fundamental question: What is man's correct relationship to the rest of nature? The technologist has one answer, the advocate for wilderness senses another. For 99 percent of the two million years on Earth, cultural man has lived as a nomadic hunter–gatherer. It is of that way of life, the most successful and enduring that man has ever achieved, that wilderness reminds us. We have learned that it was not necessarily as nasty, brutish, and short as we had supposed, and yet our interest in it invariably takes the form of nostalgia for something irretrievably lost. No one advocates a return to the "primitive." In terms of the prevailing Reason it is absurd—almost literally unthinkable—to consider it except as part of an anthropological exercise. To do so would seem to deny history. Any politician proposing a serious reevaluation of the primitive would be scorned as whimsical, and no scientist would suggest its postulation as a legitimate end of scientific endeavor. Almost all philosophical and cultural traditions stand against it. No physician could consider it for a moment, and the very demographic projections made possible by the increased control of death point to its eclipse both in nature and in human thought.

Civilization has triumphed. And yet, it has not. Ecologically our civilization is as mindless as a cancer, and we know that it will destroy itself by destroying its host. Ironically, any remnants of humanity to survive the apotheosis of civilization will be returned, genetically mutilated, to that state which we have thought contemptible. If man does not survive, "interplanetary archeologists of the future will classify our planet as one on which a very long and stable period of small-scale hunting and gathering was followed by an apparently instantaneous efflorescence of technology and society leading rapidly to extinction. 'Stratigraphically,' the origin of agriculture and thermonuclear destruction will appear as essentially simultaneous."[5]

Reason severed from instinct is a monster. It is an affirmation of intellect, therefore, and not an abrogation, to defend as a viable development from civilization a way of life in which both instincts and intelligence have flourished freely; and while wilderness is still able to suggest man's proper place and deportment, it is a narrow, hubristic, suicidal, and tyrannical Reason which will not listen.

As civilized people, wilderness preservationists have been understandably reluctant to admit this. Together with the benefits of the advanced technological society they share the fallacy of infinite expansion, or seem to do so. Radical decentralization is too anarchistic and too negative a proposal for them to make. Whenever possible they seek positive political solutions, thereby allowing themselves to enter a dialectical process by which rational "concepts" of wilderness are formulated and wilderness itself is circumscribed in thought. Should they recognize the thralldom of politics to technocracy, they will say ruefully that they are at least "buying time." But while they debate, wilderness shrinks; when they compromise, wilderness is fragmented. To endorse any projection of society's "future needs" is to endorse the growth dynamic in which technology is founded, unless the radical shift to a steady-state economy has already occurred. At the present rate of expansion, technological demand in the environment will have been multiplied by a factor of thirty-two by the year 2040, within the lifetime of children now living. It is an insane projection. Long before then we shall either have scuttled civilization, or we shall have made a reality of the Orwellian nightmare. Such words as "individual" and "wilderness" will long since have been torn from their semantic moorings.

Redefinitions are already underway. This century has seen the insinuation of the term "wilderness park" by the technocratic bureaucracy, and its ready acceptance by conservationists. In this maneuver, the State has adroitly undercut the question raised by wilderness, and has reduced all wilderness issues to the status of managerial techniques. Dangerous negative perceptions are thereby deflected into the positivistic enterprise. When the principle of management has been accepted by everyone, then the containment of wilderness will be virtually complete. There will be continuing discussions, of course, but they will be discussions among the wardens and the gardeners. No longer might the phrase "wilderness park" be seen as a contradiction in terms, for what lies within the boundaries of such parks will be wilderness by definition, and it will remain so no matter what further technological ravishment it undergoes. Wilderness hotels, wilderness railroads and airports, wilderness highways, wilderness theaters and shopping plazas—all could ultimately be made to make sense, because there will be no basis for comparison left. "Don't you see," asks one of Winston Smith's colleagues in *Nineteen Eighty-Four*, "that the whole purpose of Newspeak is to narrow the range of thought?" Should the State reserve natural areas, it will be as psychic purging grounds for those atavistic citizens who still require such treatment, but those reserves will be parks, not wilderness.

While we are able to do so, let us note the distinction. A park is a managerial unit definable in quantitative and pragmatic terms. Wilderness is unquantifiable. Its boundaries are vague or nonexistent, its contents unknown, its inhabitants elusive. The purpose of parks is to use; the earmark of wilderness is mystery. Because they serve technology, parks tend toward the predictable and static, but wilderness is infinitely burgeoning and changing because it is the matrix of life itself. When we create parks we bow to increased bureaucracy and surveillance, but when we speak for wilderness we recognize our right to fewer structures and greater freedom. Regulated and crowded, parks will eventually fragment us, as they fragment the wilderness which makes us whole.

Only when wilderness can be circumscribed in thought can it be contained, reduced, and transformed in practice. If the horizons of reason are so narrowed as to exclude radically simple alternatives, that containment can be completed. For the moment, wilderness poses its silent, subversive question. We can avoid the question. We can erase it. We can easily, most easily, lose it in a morass of technological reductions and substitutions. If we continue to act expediently, we shall at some point stand like the deracinated Winston Smith, listening to his sad song,

> Under the spreading chestnut tree
>
> I sold you and you sold me . . .

At that point our idea of wilderness will be no more than a dream of the Golden Country, a country lost forever.

NOTES

1. Jacques Ellul, *The Technological Society* (New York: Vintage Books, 1967).
2. Herbert Marcuse, *One-Dimensional Man* (Boston: Beacon Press, 1966).

3. Laurens van der Post, "A Region of Shadow," *The Listener* (August 5, 1966).
4. René Dubos, *Reason Awake* (New York: Columbia University Press, 1970).
5. Richard B. Lee and Irven DeVore, *Man the Hunter* (Chicago: Aldine Publ., 1968).

❧

"Resources Everywhere"
Wolfgang Sachs

The myopia of the conventional economists has become proverbial. While staring at the role of capital and labour, they ignore many other sources of wealth and well-being: from the unpaid labour of women backing up the world of production, to the silent workings of nature replenishing water, nutrients, and energy. Eco-developers set out to overcome this tunnel vision; they prospect the broad range of life-supporting factors to assure the sustainability of yields over the long term. Through their glasses, numerous things and actions which so far had been taken for granted as part of ordinary life acquire a new, dramatic significance: they change into valuable resources. Cow dung, for example, kindled by the Senegalese peasant to heat water in the cooking pot, suddenly becomes an energy resource; the scrap metal used by a Peruvian squatter to build an annex to his hut takes on the dignity of a recoverable input; Kenyan women cultivating village fields are discovered to be human resources for boosting food production. Under Worldwatch eyes, more and more parts of the world assume a new status, they are disembedded from their local context and redefined as resources.

In what new light, however, do actions, things, and people appear when they are redefined as "resources"? Obviously they acquire importance because they are considered useful for some higher purpose. They count not because of what they are but because of what they can become. They are stripped of their own worth in the present in order to be strip-mined for somebody else's use in the future. A resource is something that has no value until it has been made into something else. Whatever its intrinsic value, it fades away under the claim of superior interests. For more than 100 years the term "resource" has been used to survey the world for useful inputs into industry. Consequently, perception has been trained to look at forests and see lumber, at rocks and see ore, at landscapes and see real estate, at people and see human resources. To call something a "resource" means to place it under the authority of production. The old-fashioned synonym for "resources" reveals clearly how language can impart destiny: what can you do with "raw materials" except finish them in a manufacturing process? But not just any productive use can make something a resource. While the peasant in Gujarat may use cow dung to fertilize his plot, it becomes a resource only in the framework of national production. It is in national (or global) accounting books that resources are specified, measured, and assessed according to their relative produc-tivity; it is the capacity to boost GNP that constitutes a resource. Calling something a resource endows it with the availability to be exploited for the national interest.

In a non-economic perspective, things often have a meaning which makes them resistant to unlimited availability. For instance, in a Hindu village there is always a holy tree or a sacred grove which is untouchable. Gods are said to reside in their shadow; to cut them as timber would deprive the village of mighty protection. Consider another example: From Bolivia to ancient Germany, mines were regarded as wombs of Mother Earth where metals grow in slow gestation. Entering this underground world with its mysteries meant crossing a threshold into a domain which does not rightfully belong to man. Responsibility and care were required, and rituals were performed in order to ask for Mother's generosity. Cooperation of nature also had to be obtained by the North American Cree when they went hunting deer. For them, animals were not game out there to be killed, but had to be convinced, in a dialogue of rites and offer-ings, to present themselves to the hunters. Indeed, hunting was an exchange between animals and man that was governed by friendship, coercion, or love, like an ordinary human relationship. In sum, under-standing trees, rocks, or animals as animate beings in a wider cosmos where each element possesses its separate but related identity entailed intrinsic limits on exploitation.

Labeling things as "resources" takes off whatever protective identity they may have and opens them for intervention from the outside. Looking at water, soils, animals, and people in terms of resources reconstitutes them as objects for management by planners and for prizing by econo-mists. Even if they are renamed "resources" in order to maximize their efficient use, because of the cultural fall-out from the all-embracing economic cloud, it will, in the future, be much more difficult to have any intrinsic respect for them.

❧

"Bureaucracy and Wilderness (The Grand Design)"
Howie Wolke

"Foolish consistency is the hobgoblin of little minds."

—Emerson

"I'm a coniferous man, myself."

"I like just about any place that's wild," I replied, and then continued almost apologetically: "But I guess I'd agree that I probably prefer coniferous forests to deciduous."

The young forestry professor looked at me suspiciously before he continued. "Conifers are what American forestry today is all about. The old-growth deciduous forests are practically gone, and the wood products industry is now dependent upon conifers. They grow faster, are easier to manage, and are more useful to man."

The warm spring sun highlighted the forester's blond crewcut. His perfect posture and deliberate mannerisms signaled structure, order, even militarism, and he squinted through eyes that had the gleam of a man who knew he was right. We stood beneath a massive Eastern White Pine, a remnant of the mixed northern hardwood/coniferous forest that once covered most of the region. I watched the magic of photosynthesis, the fresh green five-needled bundles of chlorophyll dancing in the warm breeze. It had been a long winter, and this was a good spring day with the promise of many more to come. Just then, a truck filled with white pine logs roared past us on its way to a sawmill in the next county.

"Ah, there goes a log truck, the forester's friend." The man's face positively glowed with satisfaction.

"He's not my friend," I quickly retorted. I was now becoming annoyed, but I put a lid on it. This yahoo was my professor, a professional forester, and I was an idealistic young college freshman, already well down the road to disillusionment. I tried to ignore my rapidly developing apprehension about the career I'd wanted since my early youth, when I'd first discovered the magic of the woods, hills, and streams of central and eastern Tennessee. I still wanted to become a forest ranger.*

ON MAY 18, 1980, at 8:32 A.M., Mt. Saint Helens in Washington's Gifford Pinchot National Forest exploded. The blast, which became major world news, spewed thousands of tons of volcanic ash into the atmosphere, and in its immediate vicinity buried forests, roads, lakes, and summer cabins under burning flows of magma and layers of soot and ash. Recognizing the opportunity to preserve a newly created volcanic landscape and a chance for scientists to study primary ecological succession in such a place, Congress passed legislation, which was signed into law on August 26, 1982, establishing a 110,000-acre Mt. Saint Helens National Monument.** But the Forest Service had already been busy in its own behalf. Recognizing the inevitability of the National Monument, yet also recognizing the opportunity to prevent the Monument's boundaries from infringing upon its multiple-use domain, in late 1980 the agency began to encircle the mountain with roads and clearcuts. This, in effect, precluded Congress from including key habitat beneath the mountain—some of it undamaged old-growth forest— within the Monument. In the Forest Service, that strategy is known as "Wilderness Preventative Logging," and has been used to preclude Congressional protection of numerous areas throughout the National Forests.

---

  * Like many wilderness-oriented forestry students, I eventually became disenchanted with the timber production bias of the curriculum and the anti-wilderness sentiment of the professors. After leaving the forestry program, I began to take more courses in biology, ecology, and wildlife, finally earning a B.S. in environmental conservation with a minor in wildlife ecology.
 ** Primary ecological succession is the orderly and somewhat predictable replacement of biotic communities with other biotic communities over time on a new substrate or on one which has never supported such communities. Secondary succession occurs when existing communities are disturbed or removed and replaced with a new progression of communities.

THE FOREST SERVICE'S ANTIPATHY to wilderness is notorious. In a July 1989 article in *Forest Watch* ("Why the Forest Service Can Never Be Reformed"), National Audubon Society Vice-president Brock Evans writes:

> Once, when I was the Sierra Club's Northwest representative, I surveyed the Forest Service's record on wilderness. From 1945 to 1984, I could not find a single example of any Congressionally designated wilderness area in the four Northwest states that had been supported by the Forest Service . . . if it had stands of big trees in it that could have been logged at the time that it was being set aside.
>
> From the Bob Marshall in Montana to the Glacier Peak in Washington, from the Sawtooth in Idaho to the Kalmiopsis in Oregon—not one has boundaries recommended by the Forest Service, because they always oppose including the trees.

(It's important to note that in most cases Wilderness boundaries adopted by Congress have been only marginally better than those proposed by the Forest Service.)

Occasionally, in their frenzy to promote logging, some Forest Service bureaucrats forget to publicly feign objectivity. For instance, according to Kootenai National Forest (Montana) planner Jim Rathbun, "Cutting the suitable timber base in Glacier National Park would be for the greater social good."

The Forest Service's dedication to logging pervades the agency at all levels and in every geographic region of the country that has National Forests. For example, in Wyoming, on October 19, 1984, the Bridger-Teton National Forest's then supervisor Reid Jackson sent a letter to the chief forester of Louisiana-Pacific's lumber mill in the town of Dubois that said:

> We will continue to need your help as we work through the sensitive period ahead. All of us need to work together on gaining public support for timbering as a management tool. As mentioned during your visit, one of our primary problems in gaining public support has been, and I suspect will continue to be, a media that is biased against timber harvest where

clearcutting is involved. . . . Bob, thanks again for taking the time to visit with us. We look forward to a continued good working relationship with you and your company.

That same month, on the date of the final passage of Texas' first National Forest Wilderness bill (October 4, 1984) a field party from the Texas Committee on Natural Resources (TCONR) discovered an active timber sale in the newly designated Little Lake Creek Wilderness in the Texas National Forest. (The Davy Crockett, Sam Houston, Sabine, and Angelina National Forests have been combined under one administration.)

According to TCONR Chairman Ned Fritz:

> The purchasers had built roads into the wilderness, felled thousands of pines, and obliterated the Red Loop of the Lone Star Hiking Trail, and were still removing logs. Using heavy equipment, loggers under the supervision of the former Forest Service Ranger for the District had demolished a large fern bed, mashed the soil, and damaged the remaining hardwoods up to the bank of the creek.

It was bad enough that the Forest Service was selling timber in an area Congress was supporting for Wilderness designation (though Congress passed the legislation on October 4, President Reagan hadn't yet signed the bill); it is unconscionable that the logging operation continued, even after the Wilderness designation had become law. Furthermore, on October 15, 1984, additional timber sales were put up for bid within the Little Lake Creek Wilderness. And, beginning in January 1985, roading and logging began in the designated Upland Island Wilderness, also in the Texas National Forest. For the first time, the Forest Service was knowingly logging and roading Congressionally designated Wilderness!

But how could this be? Although loopholes in the Wilderness Act allow mining, grazing, water developments, and power corridors under some circumstances, doesn't the law at least prohibit commercial logging in "protected areas"? Wouldn't even the anti-wilderness Forest Service avoid blatantly violating such a popular law? The Wilderness logging was rationalized by the Forest Service as being compatible with section 4(d)(1) of the Act. This little-known loophole states:

... In addition, such measures may be taken as may be necessary in the control of fire, insects, and diseases, subject to such conditions as the Secretary deems desirable.

Thus, under the guise of attempting to control the Southern Pine Beetle (a native species with which the region's forests have evolved), the agency clearcut designated Wilderness. According to the Forest Service, the beetles threatened commercial timber stands outside of the "protected" areas.

Fortunately, intense public opposition, including pressure from Texas Congressman John Bryant (author of the Texas Wilderness bill), belatedly halted the Wilderness logging. Nonetheless, the Forest Service's intent was clear: establish a precedent under which logging of designated Wilderness could occur, at least in some circumstances. Although that intent failed, the Forest Service once again proved that their quest to manage, develop, and manipulate knows no bounds, from the North Woods of Minnesota to the southern pines of east Texas, and from the hardwoods of Virginia to the great conifers of the Pacific Northwest. One thing that can be said about the US Forest Service is that they are consistent. Utterly so.

Other examples abound. In 1985, Flathead National Forest (Montana) Supervisor Ed Brannon announced plans to issue a special use permit for an off-road motorcycle race in Situation 1 Grizzly habitat. Situation 1 habitat is that which is considered essential for the survival of a species listed as "threatened" or "endangered" under the Endangered Species Act. Land-use decisions are supposed to favor the listed species in Situation 1 habitat. (Griz is a "threatened" species in the lower 48 states.)

For years, the dirt bike race had proceeded each spring, eroding steep mountain soils, tearing up fragile wet meadows, and harassing various native species. Although in 1986 citizen opposition at least temporarily halted the race, the Forest Service is still pushing to allow off-road-vehicle use of the area. Indeed, the Forest Service can't seem to say "no" to any proposed use, no matter how damaging or frivolous that use might be.

The atrocities are countless. In the early 1980s, the Forest Service supported Interior Secretary James Watt's unsuccessful attempts to issue oil and gas leases in designated Wilderness, including some of the wildest and most ecologically complete units of the Wilderness System. Montana's Bob Marshall and Wyoming's Washakie were among the target areas. By then, Forest Service and BLM bureaucrats had already secretly leased about two million acres of the Bridger-Teton National Forest to oil companies for exploration and development, mostly during the late 1960s and early 1970s. The destruction of some of America's best wildlife habitat, including portions of the Salt River Range and Grayback Ridge Roadless Areas, has followed.*

Even in Indiana's tiny Hoosier National Forest, where stands of oak, hickory, beech, maple, and other hardwoods are slowly recovering from early twentieth-century clear-cutting, the logging mentality prevails. There, a 1982 analysis of demand for developed recreation concluded that the Forest's capacity for roaded recreation would exceed demand by at least 90% until beyond the year 2000. But that July, in an effort to justify additional timber sales, the Forest Service revised its estimate upward, suddenly concluding that demand for roaded recreation far exceeded capacity. According to *Forest Watch* magazine (9/87), the Forest's recreation planner explained:

> I would agree that my [July demand] calculations are high! I was told by the Forest Planning Team to make sure that demand was higher than our capability. I did as I was told.

That's corruption, pure and simple.

At the national level, in a September 9, 1983, memo to (then) Forest Service Chief Maxwell Peterson, (then) Assistant Agriculture Secretary John Crowell instructed the Forest Service to intentionally pursue a policy of violating a court decision (the law) unless private citizens actually forced the agency to comply. The Ninth Circuit Court of Appeals had upheld a previous ruling that Forest Service plans to road and develop inventoried roadless areas were illegal in lieu of an Environmental Impact Statement that complied with the requirements of the National Environmental Policy Act. Crowell's memo included this:

> I would also restate the Department's policy that timber sales and other activities not be held up or withdrawn merely because of the threat of appeal or lawsuit relying on the Ninth Circuit decision. It is important that the Forest Service put

---

* After oil companies build roads for drilling formerly wild areas, the Forest Service often "piggybacks" timber sales into the area. Without the oil/gas exploration roads, the cost of gaining access to low-quality timber would be prohibitive.

those who wish to halt development activities in the position of actually having taken the necessary steps to do so.

In other words, Forest Service policy was (and still is in some states) to break the law unless citizens force compliance by taking additional legal actions.

These examples of bureaucratic intransigence constitute the daily business of public land management in America. Although one could continue indefinitely with examples of agency corruption, disregard for the law, and single-minded dedication to development, the picture should be clear by now. And it's a sordid picture, a puzzle of sorts, composed of thousands of small pieces, many of them minor in and of themselves, yet culminatively one of the planet's major, but little-known ecological catastrophes: two million acres of wilderness devastated each year, hundreds of thousands of more miles of road planned for our National Forests, less solitude, less habitat, fewer wild animals. But more logs, more cows, more mines and oil rigs, and more work for bureaucrats. That's the grand design for America's public lands.

Throughout these pages, I have focused primarily upon National Forest issues. As I said in the Prologue, this is simply because that's where most of my experience has been. So I admit that, when evaluating America's various land managing bureaucracies, I'm biased. I'm convinced that the Forest Service is the worst. I believe that this agency is the embodiment of all that is wrong with bureaucracy, from the Kremlin to the Pentagon, and from the Bureau of Reclamation to the Bureau of Indian Affairs.

Knowledgeable persons can make similar claims about the other agencies. Some public lands defenders might note, for example, that since Congress passed the Federal Lands Policy and Management Act of 1976 (FLPMA), the BLM has gone to tremendous lengths to preclude the possibility of Congressional protection for large blocks of nearly complete wilderness ecosystems throughout its domain of generally arid and semi-arid mountains, deserts, canyons, and grasslands.*

Moreover, the National Park Service recently completed a huge new

---

* For details of some of the more glaring atrocities, I recommend that interested readers join the Southern Utah Wilderness Alliance, Box 518, Cedar City, UT 84720.

resort (Grant Village) in prime Grizzly habitat deep in the heart of Yellowstone, and, in late 1985, the US Fish and Wildlife Service bulldozed some old jeep trails into roads in its wildest unit, southern Arizona's Cabeza Prieta National Wildlife Refuge. Indeed, the quest to develop is a general disease of bureaucracy, emanating from deep within the heart of the bureaucratic beast.

The reasons for agency anti-environmentalism seem complex, and there is interagency variation; but the inner workings, problems, and prejudices of nearly all our federal and state resource managing agencies are remarkably similar. One might be tempted to simply write off the bureaucratic problem as a reflection of our industrial society. The Judeo-Christian paradigm—a central theme of which is humankind's dominion over nature—is rife within the agencies as well as within society as a whole. But public opinion polls indicate that throughout America, most individuals support wilderness; many believe in a more biocentric paradigm, at least when it comes to the management of public lands. One would expect a preponderance of these individuals in professions such as forestry, range management, and wildlife management. But instead, we discover just the opposite; particularly (but not exclusively) within the Forest Service and the BLM are a cadre of individuals who are even more dedicated to rampant industrialism than are Americans as a whole.

Public support for wilderness is obviously not manifested in the actions of the bureaucracies. In addition, while at least some land managers *know* their actions are destructive, they are afraid to act on the basis of their understanding. Job security and fringe benefits are more important to them than the Earth itself.

The reasons for bureaucratic anti-wilderness (anti-life) attitudes and actions are many. For analytical purposes, however, they can be apportioned to five basic interrelated areas, or five fundamental flaws:

FLAW #1: Each agency operates under a set of laws that are inherently biased against wilderness and the maintenance of natural diversity. The Multiple Use-Sustained Yield Act of 1960, for example, is a law under which both the Forest Service and the BLM operate. Timber, minerals, livestock, recreation, wildlife, and watershed are the commonly recognized "multiple uses" of these public lands; but one could easily add the maintenance of genetic diversity, scientific study, and education to the list, at least in the broad, if not the strictly legal sense of the term. Unfortunately, the agencies have allowed timber, minerals, and livestock to become priorities, and, when agency personnel and other pro-development interests espouse multiple use, those are the uses they represent. The doctrine of multiple use is, of course, inherently anthropocentric, and at odds with the kind of land ethic so eloquently described by Aldo Leopold in his classic, *A Sand County Almanac.**

In addition, the National Forest Management Act (NFMA) and the Federal Lands Policy and Management Act (FLPMA) basically reaffirm the anthropocentrism of multiple use. In setting procedures and guidelines for resource development, both these laws assume that development will proceed. Crucial questions, such as where might development be inappropriate, are still left to the discretion of land managers, unless Congress intervenes by specifically designating an area Wilderness or Wilderness Study.

Similarly, the National Environmental Policy Act (NEPA) requires agencies to consider a range of alternatives and their potential impacts, wherever a federal action that might have major environmental consequences is proposed. But, unless the environmental impact statement (EIS) exposes potential violations of other statutes, NEPA does not require an agency to choose the alternative of least environmental damage. Thus, once the mountains of paperwork have been completed, scores of damaging projects—oil wells, timber sales, dams, pipelines, powerlines, resorts, and roads—can and do proceed on some of the most vital and fragile public lands.

Furthermore, the 1872 Mining Law, in effect, *mandates* that when mining claims are properly established, the agencies allow the destruction of public land. And to repeat, the Wilderness Act presupposes the continuing decimation of wild America. On the whole, our public land law either encourages bureaucrats to engage in poor land-use practices, or allows them the discretion to do so.

FLAW #2: The second major problem with our land managing bureaucracies concerns the influence wielded by corporate exploiters of the public's land. The agencies are suffering a widespread identity crisis in which the *regulator* has become essentially indistinguishable from the

---

* Leopold's Land Ethic was a plea for humans to recognize the intrinsic value of all components of the biotic community, regardless of their real or perceived economic value. That ethic would be the basis for responsible land-use decisions. Leopold said: "It is inconceivable to me that an ethical relation to the land can exist without love, respect, and admiration for land, and a high regard for its value." To a growing number of public land advocates, the multiple use doctrine is incompatible with the development of a land ethic. I suggest that responsible land managers and wilderness advocates begin to think and speak in terms of intrinsic value and of multiple *benefits* instead of multiple use.

alleged *regulatee*. The timber industry largely controls the Forest Service, the livestock and mining industries virtually run the BLM ("Bureau of Livestock and Mining"), and private concessionaires have far too much influence upon the policies of the Park Service. The ways in which the timber industry governs the Forest Service range from bold to insidious.

One way is that the timber industry has embedded its fingers deeply in the pie of higher education. The conversation described at the beginning of this chapter, for example, took place in Durham, New Hampshire, on the campus of the state university. But it could have just as easily occurred in Durham, North Carolina; Syracuse, New York; Logan, Utah; Fort Collins, Colorado; Missoula, Montana, or at any university offering curriculums in forestry, range management, or other natural resource management specialties. The commodity production bias not only governs the conduct of the bureaucracies, but it also pervades the university programs which train future bureaucrats.

American forestry schools, for instance, are accredited by the Society of American Foresters, a utilization-oriented organization composed of professional foresters (many of whom work for industry) who were trained at the same institutions that they themselves now accredit. This incestuous relationship is conducive to maintaining the status quo of timber-oriented curriculums. It leaves little opportunity for the infusion of new ideas. Old dogma is simply recycled. To further complicate the problem, forestry schools receive a large percentage of their research funding from industry. Professional foresters are thus laden with training in forest economics, wood technology, "computer science," and silviculture, while their training in the natural sciences, including ecology, is generally minimal.

When I attended the University of New Hampshire forestry school in the early 1970s, forestry students weren't required to take any courses in ecology. Today, ecology and related sciences are usually incorporated into university forestry curriculums, but only as poor cousins to the more commodity-oriented subjects. For example, at the University of Montana School of Forestry (as of September, 1987), Forest Resource Management students are required to take only one 4-credit course and lab in Forest Ecology. Economics (6 credits), Forest Products and Industries (4 credits), Silvics and Silviculture (8 credits), Forest Mensuration (6 credits), Timber Stand Management (5 credits), Forest Biometrics (5 credits), and Project Design and Analysis (4 credits) still carry the day in Missoula. No wonder most young forestry graduates already conform to the production mindset by the time they enter the Forest Service!

A recent (1989) article in *The Missoulian* entitled "Timber Giant

Funds UM Slot" illustrates the corrupt relationship between universities and big industry. It reads, in part:

> Champion International Corp., a forest products company with headquarters in Stamford, Connecticut, has donated $150,000 to the University of Montana Foundation for the School of Forestry.
>
> The money is Champion's second gift to the Forestry School to establish the Champion Professor of Forestry, UM announced in a press release. The total gift for the professorship is $250,000.
>
> . . . With the donation, the Forestry School will hire a new, permanent, tenure-track faculty member. The new professor will teach and conduct research in areas of forest measurements and timber growth and yield.

The timber industry has also waged a successful propaganda war outside the universities. Slick magazine displays falsely persuade a gullible, largely urban public that non-timber values don't suffer, as long as new trees are planted on cut-over lands. Both industry and Forest Service publications (your tax dollars!) laud the brave new world of genetic "supertrees," herbicide spray programs, and other intensive management techniques, while carefully explaining that nature, on her own, doesn't always act in man's best interest. Government pamphlets (your money again) and college texts, distort biological reality by generalizing that logging benefits wildlife, even wilderness-dependent species such as Grizzlies. "Trees are a crop," say the publications; tree crops are like cornfields. If they're not harvested, they will become "decadent and overmature" (a real forestry term) and their wood will be wasted. One Forest Service pamphlet entitled "Is Nature Always Right?" tells us that "nature often works in slow ponderous rhythms which are not always efficient" and that "natural growth results in a crowded haphazard mix." To illustrate the point, the pamphlet includes a drawing of a crowded natural forest with unhappy frowning trees (frowns are drawn in the tree crowns). By contrast is a managed forest with happy, smiling uncrowded trees.

In short, the message is driven home through the news media, in colleges and universities, in government publications and documents, and by Forest Service and industry employees at various public forums. Resource managers accept half-truths as gospel, and, with impunity, twist biological wishful thinking into agency dogma. Take, for example, the common assertion that logging benefits wildlife. It is true that, in some heavily forested regions, logging—or any *natural* forest disturbance for that matter—by opening the canopy and thus allowing the growth of early successional vegetation, increases forage and habitat for some kinds of animals. Deer, moose, and some passerine birds (perching birds) are examples. But even with these seral species (species characteristically found in early successional communities), there are many exceptions.

For instance, along the Pacific Northwest coast, Blacktail Deer require, for winter forage, old-growth conifers under which snow depths remain shallow. And, in many parts of the Rockies, Shira's Moose feed extensively on Subalpine Fir, a tree of late seral stages, or the so-called climax forest.

Many species, such as most of our furbearers, various finches, woodpeckers, owls, and others, depend almost exclusively upon forests in the later stages of ecological succession, which logging destroys. Furthermore, in much of the intermountain West, timber occurs in scattered stands, and forage is generally not a limiting factor for most large mammals. Adequate cover (thick forest stands) and winter range, though, often are. Logging and associated access roads often damage both of these critical habitat components, even for species that are frequently found in early successional forest habitat. In addition, the increased access due to logging reduces populations of a variety of native animals, regardless of their natural habitat preferences. Among this group, often referred to as "wilderness-dependent species," are Elk, Grizzly, Black Bear, Mountain Goat, Lynx, Gray Wolf, Mountain Lion, Harlequin Duck, and Northern Goshawk. In an undisturbed forest, natural cataclysms such as lightning-caused wildfire, insect infestations, wind, flood, avalanche, and landslides keep enough of the forest in a variety of successional stages so that the habitat needs of early successional species are met. In many forests of the Central Rockies, Northern Rockies, and the Northern Appalachians, beavers help create a mosaic of field, swamp, pond, and forest in various stages of succession.

Furthermore, mounting evidence points to the increased nutritional quality of forage in post-fire habitats due largely to the sudden infusion of nutrient-rich ash into the soil. In contrast, clearcutting robs an area of nutrients. And, of course, logging has damaged streams and riparian vegetation throughout the public lands.

Moreover, simply due to the widespread cataclysm called *Homo sapiens,* most endangered, threatened, rare, or uncommon species depend either upon old-growth forests or an undisturbed wilderness or near-wilderness environment for their habitat needs. It is the welfare of these species—not thriving game animals such as White-tailed Deer—that should take precedence in public land-use policy and decisions. The point is that, yes, in *some* areas under *some* conditions logging—including small clearcuts—can benefit *some* kinds of animals. But that is only a very small part of the biological story. From a grain of truth (i.e., "logging benefits wildlife"), the loggers and bureaucrats have created an economically motivated lie—a distorted generalization—which has been incorporated into the forester's dogma, and which repeatedly emerges in government publications, corporate propaganda, and university curriculums. Forest industry claims notwithstanding, *logging damages habitat.* The log truck may be the forester's friend, but it certainly isn't the friend of most of the plants and animals in the forest.

Unfortunately for life, the timber industry's war against nature goes far beyond peddling propaganda; it's also a political war. The numerous big timber mills and pulp mills adjacent to National Forests throughout the country create intense pressure to maintain the status quo. Few forest rangers or local Congresspeople are likely to promote reductions in logging if those reductions might cost jobs, no matter the present level of overcutting. The logging companies maintain the threat of shutdown (job blackmail) so as to pressure the Forest Service into maintaining the levels of overcutting necessary to feed the voracious appetite of excessively large mills. In most cases, the Forest Service itself, by promising huge quantities of timber, encouraged companies to construct these mills, thus creating timber-dependent economies in small communities throughout the West. (Assistant Forest Service Chief Christopher Granger boasted in 1952 that Forest Service men were "going out and getting business.") Forest rangers generally consider it their duty to maintain this legalized insanity. Consider again the words of Reid Jackson, Supervisor (now retired) of Bridger-Teton National Forest (Jackson, WY, 6-79):

> We have a responsibility to keep the Louisiana-Pacific mill in business.

At the national level, timber industry influence and lobbying is legendary. Under the Carter Administration, timber industry lobbyists were instrumental in initiating the RARE II process (the Forest Service's second national inventory of roadless lands), which ultimately resulted in the opening of about three-quarters of the remaining National Forest roadless areas to industrial development. During Reagan's first term in office, lobbying was made easier because officials responsible for Forest Service policy were drawn from industry. For example, John Crowell, previously the chief legal counsel for Louisiana-Pacific (the largest purchaser of National Forest timber) was the assistant secretary of Agriculture in charge of the Forest Service. During most of the Reagan years, both the agency's Chief, Max Peterson, and its Deputy Chief, Jeff Sirmon, were professional road engineers.

Further exacerbating Forest mismanagement, timber production "targets" for each Forest come directly from the Chief's office in Washington, D.C., far from the land, but close, of course, to Washington's cadre of timber industry lobbyists. As we shall see, career advancement in the Forest Service is for those supervisors and district rangers who meet their predesignated timber volume, road mileage, or other targets. Therefore, the designation of Wilderness can easily be perceived by agency employees as a direct threat to career advancement.

In sum, through a well-organized multi-level propaganda campaign and through the application of tremendous political pressure at the local, regional, and national levels, the timber industry has cleverly become so intertwined with the US Forest Service that it is now difficult to distinguish between the two entities. At this point, it might be tempting to simply attribute bureaucratic anti-environmentalism to ineffective public land laws (ineffective for wildland advocates, that is), and the strong-arm tactics of timber, mining, grazing, and resort industries. But these factors are only a part of the story. Far be it from me to attempt to absolve a gutless, immoral, or unresponsive Congress, or a greedy corporate America for their roles in wildland decimation. But classic liberal dogma is oversimplified. The answers to our public land deprivations do not lie merely with the passage of stronger laws to control corporate exploitation. The result of more regulation in the absence of more substantive changes in bureaucratic (and public) attitudes and structure would not be, as many liberal-leftists would have us believe, benevolent bureaucrats acting in the public or ecological interest.

In America, society as a whole expects companies to promote self-interest, even when corporate actions directly oppose the welfare of the general public or the environment. But society also expects public land managers to regulate the actions of exploitative industry. Unfortunately, bureaucracies such as the Forest Service and the BLM promote destructive exploitation while pretending to act in the public and ecological interest, and that makes their crimes against nature far more atrocious than those of the exploitative industries. Though many agency decision-makers no doubt believe their destructive actions to be proper (that is, to be in the nation's economic interest), it is the results of their actions, not their motives, that count. And again, the results are the devastation of our public lands. Corporate America continues to play its role effectively, but the agencies have completely abrogated their responsibilities to both the land and the public.

The gradual shift of the agencies from regulators to promoters is indicative of the depth of the bureaucratic problem. The bureaucrat-sponsored destruction of wilderness continues because of inherent flaws in the bureaucratic system—flaws that cannot be rectified merely by passing more laws or instituting more regulations. I'm convinced that, if timber industry lobbying were outlawed, if livestock industry influence on the BLM somehow was negated, if National Park concessionaires were nationalized, the agencies would continue to hack, gouge, and scrape away at our public lands, albeit not quite so rapidly as they do now, but as overseers of the rape of wild America nevertheless.

In addition to the problems of faulty laws and corporate influence (Flaws 1 and 2), there are three other fundamental causes of bureaucratic anti-wilderness actions and attitudes. These last three reasons pertain to the structural problems within the bureaucracy itself and within its system of incentives.

FLAW #3: The third major reason for widespread bureaucratic anti-environmentalism is that agency decision-makers tend to be the survivors of an incredibly efficient Orwellian "filtration system." The structure and promotional policies of land managing agencies generally do not allow reform-minded or wilderness-oriented professionals to advance into decision-making positions within the bureaucracy.

The beginning of the filtration process occurs at the university level. Students who appreciate wildness, and whose biocentric viewpoints have led them to enroll in natural resource curriculums, tend to become rapidly disillusioned by timber-oriented forestry programs, livestock-oriented range programs, or hunter management-oriented wildlife programs. Many either drop out or switch programs, leaving the natural resource profession with a preponderance of unquestioning, multiple-use-oriented graduates ready to be employed by the agencies. Few young people with vision, ideals, courage, and a love of raw nature ever enter the resource management professions. Most have already been filtered.

Federal employment usually means good pay, superb benefits, and, above all, job security. "Don't rock the boat" and "be a team player" are the battle cries within the Forest Service, BLM, US Fish and Wildlife Service, and National Park Service. Those few advocates of sane public policy who do become established in the lower echelons of the agency—and there are some good timber foresters, wildlife biologists, soil scientists, range conservationists, hydrologists, and recreation specialists—find career advancement to decision-making positions difficult or impossible. At some point, they're forced to choose between their careers and their principles. Few have the courage to publicly oppose agency policy; those who do face transfer and career stagnation.

In the Forest Service, the quota system is a major factor in promotion policy. The district ranger who meets his timber or road mileage quota is likely to become a Forest supervisor. Likewise, the supervisor whose Forest meets its timber quota, especially if he can do it with minimal controversy, is likely to become a regional forester.

In short, subscribers to agency dogma advance, while the few renegade advocates of responsible government stagnate in the bureaucratic fuel filter of the industrial machine. In the bizarre world of officialdom, it is the conformist, not the innovator, who climbs the ladder of individual success. Tradition has degenerated into a self-perpetuating cycle of destructive conformity. Like scum on the surface of a sewage lagoon, the worst of the bureaucrats inevitably rise to the top.

FLAW #4: In the Forest Service, there's another reason for agency anti-environmentalism that is closely related to those already mentioned. Simply stated, the more timber that is sold on a particular National Forest, the more money the Forest receives for various management activities. There is an *economic incentive* for forest rangers to produce sawlogs. Under the Knutson-Vandenberg (K-V) Act of 1930, the Forest Service may retain a portion of its timber sale receipts for a variety of management activities, including reforestation, brush disposal, timber salvage sales, wildlife habitat "mitigation," and administrative overhead. Prior to a timber sale, the Forest Service prepares a K-V Collection Plan which specifies how it hopes to use the receipts. Under the National Forest Management Act, each Forest must insure that the federal treasury receives a minimum of 50 cents per thousand board feet of timber sold. This amounts to a tiny sum.

Accordingly, if a particular timber sale goes for a high price, the FS might cover its entire K-V plan with receipts from that sale, *and* return a substantial sum of money (beyond the 50 cents per thousand) to the US Treasury. But, if the timber is less valuable, K-V projects will utilize all or nearly all of the receipts, resulting in the treasury receiving only the 50 cents per thousand board feet, or perhaps a bit more. When timber sale receipts are very low, direct appropriations must pay for the projects that the K-V funds were not sufficient to cover.

Most National Forest timber is of low commercial value, and sales frequently return the minimum to the US Treasury. This is especially true in the cold dry Rockies where trees are small and slow-growing. In the more hospitable climates of the deep South and Pacific Northwest, however, National Forest timber sales often return substantial sums to the Treasury.

Indeed, when all timber sale related costs for road-building, reforestation, wildlife habitat mitigation, brush disposal, and administrative overhead are calculated, most National Forest timber programs (except in the South and Northwest) result in big net monetary losses, *even though the individual forests can increase their budgets* (budget maximization) *via K-V receipts by selling low value federal timber.* Thus, foresters can fund their favorite programs, create projects, and increase their own and their district's prestige within the agency by selling timber. That's one reason why in the National Forests in 1985, below-cost timber sales cost American taxpayers at least 600 million dollars (according to the OMB).

Before describing Flaw #5, it is worth noting the means by which most roads in National Forests are funded: Logging roads are subsidized by the taxpayer in two major ways. First, direct Congressional appropriations fund most of the major haul roads that the Forest Service bulldozes into previously unroaded terrain. Second, purchaser credits—an indirect subsidy—generally fund the shorter spur roads used to gain access to particular timber sales. To finance these "Purchaser Credit Roads," the FS subtracts the estimated cost of road construction from the stumpage price at which bidding begins, in exchange for the logging company doing the actual road-building. Without the road subsidies, direct and indirect, many destructive timber sales simply wouldn't fly, because road construction costs would exceed the worth of the timber. Road subsidies are a key reason why the Forest Service loses so much money selling trees.

FLAW #5: The fifth major reason for agency anti-wilderness inclinations is perhaps the all-pervasive consideration to which the first four are subservient. It is the fundamental flaw in the bureaucratic system, and it is the flaw that—even if there were serious reforms in environmental law, industry influence peddling, university programs, and agency promotional and funding policies—would perpetuate abuse of our public lands. Simply stated, wilderness reduces the need for bureaucracy. To advocate wildland preservation is to advocate fewer bureaucrats. Thus the despicable behavior of bureaucrats is inherent to the bureaucratic system. The present system is not likely to be reformed in any meaningful way; it is rotten to the core.

The professions most relevant to public land administration are often referred to as resource *management* professions. Foresters, range specialists, and other land *managers* are not in the business of just sitting back and watching natural processes occur on the public lands. After all, those lands are not there just to photosynthesize, transpire, and respire. They must be *managed*; they must be used. The bureaucrat has been trained to *manage*, and has invested large sums of money and at least four years of higher education to pursue a career in *managing* resources. You cannot tell a resource manager to cease managing resources any more than you can tell an artist to cease painting. The urge to manage is deeply ingrained within the beast. When there is timber to be sold, a road to be built, an oil well to be drilled . . . plans must be formulated and carried out, and so-called "mitigation" measures must be conceived and implemented. All this requires planners, timber foresters, recreation specialists, mineral specialists, engineers, hydrologists, soil scientists, wildlife biologists, and other members of the federal work-force. Granted, the employment of these specialists on many projects reduces ecological damage, but most often the best way to avoid the damage is simply avoid the projects. But that's not what these public servants are paid to do.

The formal land-use planning process requires tremendous inputs of time and labor. For example, the National Forest Management Act of 1976 requires the Forest Service to produce a planning document for each Forest every 50 years, with updates at 10- to 15-year intervals. The documents are commonly over a thousand pages in length, and each represents thousands of work hours by supervisors, district rangers, and various resource specialists often referred to as "I.D. Team" (interdisciplinary team) members. Most of these plans promote habitat destruction through the massive infusion of energy and funds for road-building, logging, and other management activities (remember, the Forest Service now plans a minimum of 75,000 miles of new roads in remaining roadless areas by the end of the century). The perpetuation of bureaucracy and the destruction of wild country are the cause and effect Siamese Twins of public land management in America.

The bottom line is that intensive multiple-use management keeps the bread and butter on the bureaucratic table. Of course, roadless areas and designated Wilderness Areas also are managed (trail maintenance, rehabilitation of overused sites, user education, outfitter permits, fire management, etc.), but it is a far more benign form of management, and requires far fewer resource managers than traditional multiple-use management. To advocate Wilderness, or even just more benevolent forms of multiple-use management, is to threaten not only their career advancement, but ultimately the very existence of individual bureaucrats. When we recognize the innate contradiction between bureaucracy and wilderness preservation, then we realize that it is naive to expect land managers to advocate programs not requiring much management, and that there is a built-in limit to the usefulness of conservationist dialogue or negotiation with agency bureaucrats. I do not suggest that wildland defenders eschew negotiation in all instances. But there are deep-rooted reasons why land managers usually side with developers. The bureaucratic system rewards self-interest and agency loyalty, which are the basis for many poor land-use decisions. The major function of the bureaucracy is to perpetuate the bureaucracy.

Other reasons could be cited for the developmental fanaticism of federal bureaucrats. Furthermore, the five fundamental reasons we've examined are all interrelated and mutually reinforcing. For example, a citizen proposal to the Forest Service for protecting a particular area from logging is likely to fall upon deaf ears, because protection would make it more difficult for the ranger to meet his timber quota (thus threatening his career advancement), because of the hostility the local timber industry would direct toward him, because selling timber increases his Forest's budget, and, in a deeper sense, because proposals to maintain wild areas ultimately represent a threat to the very need for bureaucrats. Also, after four or more years in forestry or related resource management studies, and years within the agency surrounded by agency dogma, decision-makers probably *believe* in what they are doing, perhaps to the same extent that wilderness advocates believe in their causes.

The bureaucratic problem gives us reason to doubt that young foresters, range managers, or other agency decision-makers are likely or able to initiate major reforms. Even though the US Forest Service, for example, was a utilization-oriented agency even at its inception under Teddy Roosevelt in 1906, its founders would be appalled at the agency's holy war against wilderness, its massive-scale road-building and clear-cutting practices, its mismanagement of livestock grazing, and its general callousness toward wildlife, water quality, and soils. The true utilitarians of the early 1900s, such as Gifford Pinchot, as well as the early biocentric visionaries, such as Muir, Leopold, and Marshall, would view the Forest Service of the 1980s with absolute revulsion.

During my first ten years as a wilderness activist in Wyoming (I'm a slow learner), I often participated in, and indeed sometimes initiated, negotiating sessions with FS and BLM bureaucrats. The usual focus was a proposed timber sale, oil rig, road project, land-use plan, or, occasionally, a citizen proposal to manage a sensitive area in a roadless manner. During this period, I saw no reason why these people wouldn't be open to moderate proposals and logical arguments based upon sound biological fundamentals. I assumed that, in spite of our differences, these people must have some feel for the land since most of them presumably had originally pursued a natural resource career because of a love of the outdoors. Many agency personnel, moreover, are personable and seem reasonable on a one-to-one basis. Indeed, most of the meetings were positively cordial. But I hadn't considered the five flaws of bureaucracy. During this period, *not once* did I see a Forest Service or BLM manager voluntarily reverse a major anti-wilderness decision as the result of a dialogue or negotiation. *Not once* have I seen an agency decision-maker voluntarily put the public interest ahead of bureaucratic or personal considerations, though I have seen lower level agency personnel attempt to do so. And *not once* have I seen a major environmental gain that wasn't the direct result of legislation, litigation, administrative appeal (including the threat of legal action), or intense grassroots opposition to agency policy.

Here in the Rockies we have managed to stop some especially bad timber sales, some road-building projects, development of a few particularly egregious oil and gas projects, and to include some major areas in the National Wilderness Preservation System. But the victories have almost always been over the kicking and screaming objections of FS or BLM bureaucrats, whose original mission, we must keep reminding ourselves, was to conserve the nation's limited natural resources and to protect the public, not the corporate interest. It would behoove the wilderness movement to take note of the reasons for bureaucratic intransigence, and to formulate tactics and strategies based upon the incontrovertible evidence that bureaucracies in America today cannot be reformed without major structural overhaul, and cannot be expected to voluntarily promote ecologically sane land use in the absence of strife, conflict, and confrontation.

From
A Wilderness Bill of Rights
William O. Douglas

Our ethic has become the automobile, the bulldozer, the industrial plant. The growth factor in gross national product is the controller before which all must give way. The meadow, the swamp, the wooded alcove and their inhabitants must surrender. Commercial and mechanical recreational use and productive use come first; conservation use is low on the totem pole.

We need a new conservation ethic if we are to have sanctuaries of wilderness left commensurate with the need. This ethic was described by Leopold in *A Sand County Almanac:* "A thing is right when it tends to preserve the integrity, stability, and beauty of the biotic community. It is wrong when it tends otherwise."

This means education, starting in the early years, which introduces reverence, wonder, and awe of nature, not the power of the bulldozer, as symbols of virtue. It means introducing biologists, botanists, and ornithologists, as well as engineers and landscape architects, into our planning councils. It means the invention of new procedures that hold the hand of the developer until all alternatives to defacing rolling green hills or destroying wilderness sanctuaries are exhausted. And it means above all else making conservation a civic cause and uniting all of its advocates in a united front against the hundreds of threatened invasions that appear year after year.

✿

"Travels with Seldom"

Ellen Meloy

On a clear May morning, ten people board two boats near Utah's Bullfrog Marina and roar down the "Blue Death," as writer Edward Abbey called Lake Powell, to a base camp near the mouth of the Escalante River. From this camp we will explore side canyons on foot for four days. Veteran Utah guide Ken Sleight leads the trip, and it is an anomaly. Most of the boaters we pass are engaged in strenuous water sports: microwaving pizza, rocking out to Dirt Love Injection tapes, raking astroturf decks, sunbathing atop the ubiquitous houseboats that chug down the lake.

We sit several hundred feet above the Colorado River bed in Glen Canyon, which Sleight first ran in a fat-tubed rubber military assault barge nearly forty years before. What holds us up today are the powerboats and Lake Powell, the vast reservoir that extends two hundred miles into Arizona and Utah behind the Bureau of Reclamation's Glen Canyon Dam.

A few times during the ride to base camp Sleight has called Lake Powell the Blue Death or Lake Foul, names popularized by Abbey. Mostly he calls Lake Powell "the River," as if this inland sea of perpetual houseboat gridlock, visited by three million people a year, simply didn't exist. For him this is trip number who-knows-which; he can't remember, so many times has he traveled this piece of Colorado Plateau. Despite the lake atop the river; despite the personal ordeals of watching Glen Canyon go under and the death in 1989 of his friend Abbey—who Sleight believed gave voice to his own rage and grief and resistance against man's abuses of Utah's slickrock wilderness—no matter that the world has changed, Ken Sleight is still "running" Glen Canyon.

Today's pilgrimage might seem disturbing for someone who watched a river die. Instead Sleight sits in the boat grinning as if he were just caught in the attic reading cowboy novels. "I hate Lake Foul," he laughs. "But I like to come here once in a while. It makes me go back home and fight."

Sleight was the model for Seldom Seen Smith, one of the quartet of eco-guerillas in Abbey's 1975 novel *The Monkey Wrench Gang,* and its sequel, *Hayduke Lives!* Proprietor of Back of Beyond Expeditions, a jack Mormon, or apostate, but "polygamous as a rabbit," Smith was "a lanky man, lean as a rake, awkward to handle," Abbey wrote, with "a nose like a beak, a big Adam's apple, ears like the handles on a jug, sun-bleached hair like a rat's nest and a wide and generous grin." The description still fits, though Sleight is in his sixties now and has married only twice, one wife at a time. When he is not guiding, he devotes a great deal of time to environmental activism and the southern Utah guest ranch he operates with his wife, Jane. The rat's nest is gray, his hands are young and gentle, his aura of irrepressible mischief remains intact if not raised to new heights.

For me this is trip number one on the Blue Death. I was in Mickey Mouse pajamas when the hydrobinge on western rivers peaked with Glen Canyon and Flaming Gorge Dams. I knew that for conservationists Glen Canyon Dam was an act of gross vandalism. For others it is a monument to progress and human ingenuity, a recreation paradise, an economic boon, and for its energy supply, a necessity. Until this trip, however, I turned my back to Lake Powell. When there was no choice but to take the roads that skirted it, I hunched my neck, ground my fillings, riveted my eyes to the centerline, and drove around it like a crazed walnut. One exception occurred after a hide-scorching hike near Hite, a marina on Powell's upper reaches, when I waded through log debris and nervous flotillas of Styrofoam to bathe in the lake. I was joined in my swim by a large red sofa.

Sooner or later everyone must come to terms with Lake Powell's duality as hubris or techno-triumph, so I'm here to dip more than toes into the Blue Death, to travel with Sleight, who remembers predam Glen Canyon better than anyone else.

I SURVIVE MY FIRST NIGHT on Lake Foul. As we motor to Rainbow Bridge, where we will hike today, I am in fact astounded by my coldly dispassionate attitude. Look at undammed Glen Canyon and Lake Powell through the different ends of binoculars, I think as we slurp over the wakes of houseboat fleets. So vast a maze of bare rock and wild river was Glen Canyon, so aloof and remote, the long-range view was far too heady. Better to pull in the details, the silt-gold river surface, crimson blossoms of a claret cup cactus, a lizard's turquoise throat, the mossy grottos, or glens, that gave the canyon its name. The beauty of Lake Powell is best seen from the end of the binoculars that distances the view: stunning azure water with a shoreline labyrinth of sensuous red, beige, and ochre sandstone domes, a lake many people call the most beautiful in the world. Up close it is a flooded river gorge lined with tons of Gulf of California—bound but dam-arrested mud, cans, bottles, batteries, stolen cars, six billion pairs of sunglasses.

Sleight and another trip companion, Bill Adams, a professional magician from California, do not have to warp their perception. They saw Glen Canyon as it was. Everywhere they look as we motor down the lake to our trailhead, they can peel back this water and remember. It is said that you can see a person's true face only in the glare of lightning. When I watch these two men remembering Glen, it is as if in lightning.

Over thirty years ago Adams shopped for outfitters in southern Utah. "I got a lot of slick brochures," he recalls. "Ken sent me a mimeo on beige paper. The lines ran together, the ink was smudged. I signed up." It was his first of ninety trips.

Sleight's guiding career began in the mountains near his boyhood home in Paris, Idaho, a farm town on the Utah border. Son of Mormon pioneers, he was the kid who rounded up friends and instigated the hikes and overnights. "I led the pack, just like I do now," he says sweetly, and you know you would follow him, too. He served in Korea, earned a business degree, married, started a family, taught school, and in 1955 bought a few forty-dollar military inflatables, in which he carried Boy Scouts, youth groups, the stray flock of little old ladies through Glen Canyon. As his clientele broadened, he worked other stretches of the Colorado and

Lodore and Desolation Canyons on the Green River. Off season he outfitted horsepack trips in the Escalante backcountry.

When river recreation boomed in the late 1960s, outfitters added more boats, hired more boatmen, cranked more people downriver. Federal river managers invoked a permit system giving outfitters and private boaters pieces of a pie, precariously balancing intense demand with the river's carrying capacity. Sleight kept his operation small, essentially one-man. He lobbied for a ceiling on the size of river companies, arguing that they should remain family-sized operations—businesses, not industries—suitable to their medium: the River.

For his lifelong defense of small business, Sleight has been called an anachronism and crazy. Crazy to cap the growth potential of his own enterprise, to resist the inevitable, which indeed came to pass. Commercial river running is a competitive, largely standardized recreation industry. Roughly twenty companies now run each stretch of southwestern whitewater, each carrying as many as twenty-five passengers on brief trips that maximize turnaround and profit.

Sleight turned his river business over to his son Mark in the early 1980s but continued to run Glen. "Someone has to remind the world that back when Reclamation built the dam, they promised that Lake Powell would forever remain the domain of small businesses," he asserts. Within less than twenty years of that promise, however, the Del Webb corporation, followed by ARA Leisure Services, controlled all marinas, accommodations, and tour and boating services, a justified evolution, according to the National Park Service, which manages Glen Canyon National Recreation Area, since a single concessioner has the capital for development and can handle great numbers of people more efficiently. Several years ago the Park Service tried but failed to revoke Sleight's Glen Canyon permit on the grounds that his outfit was "too small to serve the public interest." "I enjoy being all that stands between ARA and a monopoly on Lake Powell," he says, as if he were talking about a rather pleasant game of golf.

In the summer of 1962, Glen Canyon was terminal. The river outfitters announced their "Farewell Voyages" in the July issue of *Desert* magazine, the requiem pilgrimages down a river soon to turn slack behind 800,000 tons of concrete countersunk between sheer walls of Navajo sandstone. Ken Sleight Expeditions was among the listings: group size, twelve; cost, one hundred dollars per person.

The Sierra Club and other conservation groups had resigned themselves to what they called "responsible reclamation," and much to their later horror, they virtually let Glen Canyon go. Conservation's big battle was to the north, against dams in Dinosaur National Monument. At Glen Canyon the case for mass recreation for "the common man" touted by the irascible Floyd Dominy prevailed. A pristine, inaccessible river was worthless, the developers argued, sounding a refrain that haunts environmentalists to this day: Those who want to embalm nature are elitists. "Dear God," prayed Dominy in a book about Lake Powell, "did you cast down two hundred miles of canyon and mark: 'For poets only'? Multitudes hunger for a lake in the sun." Dominy's masses cameth and they are still ravenous. The waiting list for a houseboat rental on Lake Powell can be two years long.

Even without the big conservation groups, Friends of Glen Canyon, a loose assembly of boatmen and outfitters, including Sleight, made a case for damning the dam. "We hardly knew what we were doing," he remembers. "We didn't have documents to back us, we didn't know how to tell other people about the desert wilderness that would be lost. I spoke out, I said, 'This is what this place means to me.' But every politician, bureaucrat, and developer in the country wanted to dam it, and they did."

Sleight ran Glen when the dam gates closed in 1963 and the river went slack. He added motors and ran it some more, watching the reservoir rise "inch by lousy inch," an embittered witness to the disappearance of over a hundred canyons, thousands of Anasazi Indian sites, graceful sandbars, alcoves large enough to hold a symphony orchestra, grottos of gentle seeps and lush greenery, Music Temple, Cathedral in the Desert, Hidden Passage, cottonwoods, willows, saltbush, herons, snowy egrets, coyotes, deer, and millions and millions of grains of exquisite red sand. The slow inundation was a tragic personal ordeal. "I felt like I was in a straitjacket," Sleight says. "The world fell down around me and I couldn't do anything about it."

From a boat dock with floating rest rooms and a sizable pod of Dorito-addicted carp, we walk up a paved trail to Rainbow Bridge. Behind us the *Canyon Odyssey,* a sleek white multidecked tour vessel operated by ARA, pulls into the narrows, its engines idling like a lineup of 170-horsepower Cuisinarts on PUREE. "Please return to the boat in twenty minutes," commands a robo-voice over a loudspeaker, loud enough to tremble the prayer feathers atop nearby Navajo Mountain, and a large crowd shuffles obediently forward in their aloha shirts to look at the nearly three-hundred-foot-high, forty-foot-thick arc of sandstone. *Nonne'zoshi,* in Navajo. The stone rainbow.

"It was a six-mile hike from the river to the bridge," Sleight tells me. "A long, hot hike. Some people, before they turned the last bend and saw the bridge, were ready to strangle me. We left cans under the seeps in the canyon walls to collect water for the next batch of hikers. The creek was cool and clear, with pour-offs and pools. Redbud trees over there"—he points to the banks—"a hanging garden in each alcove. Rainbow Bridge was a monument to wild country, to solitude."

The 1956 law creating the Upper Colorado River Project contained a provision to protect Rainbow Bridge from encroachment or impairment by reservoirs. In a seven-year lawsuit, Sleight, Friends of the Earth, and the Wasatch Mountain Club fought to keep the government from breaking its own law and letting Lake Powell reach the bridge. A district judge ruled to limit the lake's height, but the Supreme Court overturned the decision, and the water crept up the canyon walls. In most years a blue tongue of Lake Powell covers the creekbed and laps the buttresses of the stone rainbow. In these recent years of low lake levels, the tongue has vanished, leaving a broad swath of lifeless, gray-brown sediment.

"We used to climb up the dome of rock on the right, then lower ourselves off a rope to the buttress and onto the top of the bridge." Sleight pauses, grins. "I think I'll do that again someday, when the rangers are here."

More tourists spill down the trail that makes Rainbow Bridge not the secluded domain of desert rats and semidehydrated Boy Scouts, but truly accessible to the motorized hundred thousand visitors per annum who won't get blisters or strangle their guide. Looking impish in his faded cowboy snap shirt, Sleight walks up to a woman in the crowd. "Want to go up there?" he asks, pointing to the top of the bridge. "I'll take you there for a hundred dollars."

After years among river people, the ones with perpetually sandcaked motleyness and a look that can be read either as supreme enlightenment or extreme horniness, I am still amazed that we call this organism "the River" when it is actually several rivers of the Colorado River system, numerous segments with no resemblance to a river, and a few more with no water in them whatsoever. Each time we say "the River" we seem to resurrect the lost wild country.

Sleight is a more proactive resurrectionist. In 1988 several hundred people gathered on the crest of Glen Canyon Dam to sing the dam's praises in its twenty-fifth year. Sleight was there, and as he walked toward the speaker's podium to present a statement with a different point of view, security guards trailed him as if he were Muammar Qaddafi strolling through the lobby of the Tel Aviv Hilton. Sleight won't let go. He simply believes there is no end to the dam's story until the Colorado River breaks loose.

Helpmeet in his battles, albeit posthumously now, is Edward Abbey, whom Sleight met in the seventies at Lee's Ferry, Arizona, when Sleight was rigging a trip down the Grand Canyon. They talked into the night, Sleight recalls, mostly about Glen Canyon Dam. In what would become a lifelong discussion, wishful thinking was epic. Some of it found its way into *The Monkey Wrench Gang*. Dynamite. ("We get three jumbo-sized houseboats and some dolphins. . . .") Earthquakes. (Seldom Seen kneels atop the dam and prays, "God. How about a little old *pre*-cision-type earthquake right here under this dam?") Levitation. (Abbey, Sleight, and others once circled a campfire, grabbed hands, shut their eyes, and tried to raise the dam, just a few inches, to let the water out.)

Abbey went on many Sleight trips over the years. The two "wild preservatives" had much common ground: belief in open government, social and environmental justice, the need to dog destructive developers and the public servants who forget whom they serve. Both were hard in their vision of a restored Glen Canyon. "Lake Powell will disappear," Abbey wrote in 1988. "Del Webb will go back to Lubbock, Texas, where he belongs."

In camp I eat breakfast alone on a shoreline ledge. Above it rises the sandy slope of camp, then a sheer eighty-foot cliff. Below this ledge, under Lake Powell, is the rest of the cliff and the old river bottom, silent in its aquarian tomb. A canyon wren sings its glissade of tender notes, and in the soft morning light the lake is lovely, not counting the thick, dirty-sock white crust of calcium carbonate directly above the water's edge: Lake Powell's bathtub ring, sixty feet high in this interminable drought. Suddenly I am weeping. I cry because I cannot smell the heady musk of cottonwoods or feel the quiet movements of water in the unlikely ribbon of paradise that curls through acres and acres of bare, baking slickrock. I cry because three million people a year love the Blue Death and I don't. I weep for Glen Canyon and I have never seen it. I am not the one who sunk his life into this place—"the heart of his river, the river of his heart," as Abbey called it. He's over there greeting each guest.

"Good morning! I see you lived another day," Sleight says to the family from Maryland.

Tears drench my shirt, my hair dangles in my salsa, a hot pain rises with such force, it blurs my vision, knocks me over, and as I hit the rock my heart splits with a loud, painful crack. I sob and mumble about dolphins and explosives.

"I sure hate to see people die on my trips," he teases. "It's such a nuisance taking them out. You have to tie them to a mule."

Later I am cheered by Bill Adams's repertoire of magic tricks and a tale of the time when, before a trip down the Grand Canyon, one of Sleight's old neoprene rafts blew up, an event forever known as "a Sleight explosion." I relish our final day of hiking, and I try hard to fill my coward's heart with the hope of Sleight, Abbey, and others for whom Glen Canyon was a tragedy but also a force around which to rally and say, No more wasting of places like Glen Canyon, no more oversized, outstripped, ultra-zealous transcendence of the West's own limits, no more trying to make Utah, Arizona, New Mexico, and the rest of the desert look like upstate New York.

On our last night in camp, Sleight is laid-back, gently attentive to each guest, and speaking gross understatement. "Sure, it hurts to come back to Glen," he tells me. "I was lucky to see it, but there are lots of canyons I haven't seen." Lavender shadows etch the canyon's vermilion walls and deepen in the dusk. We recline against duffel bags on the quiet beach.

I suggest a run to Vegas for explosives. I imagine a bunch of people with terminal diseases, flinging themselves off the dam's lip with dynamite strapped to their backs, their shoes on fire. The stories of how to "do" the dam are, like the sheer catharsis they serve, epic and legendary.

For years Sleight has written letters, spoken at hearings, insisted that all sides be heard, acted to set right a wrong after everyone else gives up. While he supports environmental groups, he prefers to fight solo—Don Quixote, a friend calls him. Not long before this trip, he had ridden his horse in front of a bulldozer at a four-hundred-acre Bureau of Land Management chaining operation on a mesa near his home. (Chaining is when two bulldozers drag a ship's anchor chain between them and grind across the desert, uprooting century-old pinyon and juniper trees and all other vegetation in their path. The land is then reseeded for cattle and wildlife forage.) All avenues of public protest against chaining on the mesa had been tried. Still, Sleight felt it was wrong. He blocked one of the bulldozers, dismounted, and asked that the operation be stopped. He was wrestled aside. The bulldozers turned the desert upside down.

Sleight's style tends to mythologize him, to make him more Seldom Seen than a man whose grip on his monkey wrench is actually quite simple: unfailing truth to himself. To Utah's canyon country, to freedom, to the River, to knowing a place intimately and leading a small group of explorers into its heart, to keeping one's work well within the reach of love.

On the beach we talk of his foray into politics. He ran for a seat in the Utah legislature, lost the race, but still dreamed of the day he might ride his horse up Salt Lake City's capitol steps, followed by packhorses bearing briefcases, and "fix" a few laws. Again, humor defuses the seriousness of a tough, consuming personal endeavor.

We talk of Ed Abbey.

"He is supposed to be up there arranging that *pre*-cision earthquake." Sleight laughs. He tells me the story of the time, in Cataract Canyon, he was thrown from rowing position and yelled for Abbey to take the oars before they dropped into a maelstrom of whitewater known as Satan's Gut. "Pull, Ed, pull!" Sleight cried. Abbey pulled, there was a loud crack. The oar splintered. Abbey looked into the frothing hole, then at Sleight. "Ken," he said with a silly grin, "do we have another oar? This one seems to have an imperfection."

"With Abbey I always felt we could solve problems, we could set the world right," Sleight says. Our lakeshore camp is dark, everyone else is asleep, but we're still sitting in the sand. "On every trip together we'd try to figure out just how to do it.

"His pen was his monkey wrench. We agreed on so much, never have I felt such a kinship with another person. All my life I couldn't seem to say what I wanted to say. Abbey said it for me. He was my voice, our voice. I'm barely getting over losing him."

"Dear Ed," Sleight began in a moving letter to his friend, read during Abbey's memorial at Sleight's ranch. "Dear Ed, I wish I could rise above this."

I bid Ken good night and walk across the starlit beach to my sleeping bag. The canyons and the entombed river beside me slipped away before anyone could comprehend their wild, wild souls. Pioneers and explorers like John Wesley Powell ogled this place, never quite believed what they saw. Sleight, Abbey, and their contemporaries tasted it, its flavor barely on their tongues. Too many among those born after the "wild preservatives" do not even know that a river once flowed here. This lake, they believe, came along with everything else in the Book of Genesis.

Suddenly I remember that nearly thirty years ago this very day, Glen Canyon Dam shut off the Colorado River. I crawl into my bed by the Blue Death. I make plans to sleep tomorrow by the River, the part that still flows by and makes sounds. I try to rise above this.

❧

"The Stone Gallery"

Bruce Berger

Slickrock that leaps unexpectedly into a dome, an arch, a spire, causes a
similar leap in the hiker's pulse. Stone and erosion, working their uniform
processes, seem newly charged with the miraculous, and the dullest sandstone is suddenly potent with twists and acrobatics on its journey back to
sand. One learns again that unexpectedness is the wellspring of beauty.

But name the phenomenon Druid Arch, Yellowstone Falls, Zabriskie
Point, Capitol Lake or Devil's Tower, and put it in Arches National Park,
Death Valley National Monument, or the Maroon-Snowmass Wilderness
Area. Suddenly the object detaches itself from its context, gains an invisible frame and becomes a classic. Gone is the weave of intricate forces; one
senses instead the hand of the individual. The tide of nature is parted for
the revelation of some miracle, and behind the miracle is the Word with
its single, limiting vision. For reasons lost in the matrix of language, the
christening of a piece of nature transforms it into a work of art.

Our national parks and monuments, promising continuity, respite,
and pre-civilized experience have increasingly become mazes of human
language. One goes there to regain some nameless belonging, only to
encounter a catalogued gallery. Displays at park headquarters instruct one
in proper appreciation. Roads, trails, and signs lead from one novelty to
another. And the scenic loops are studded with statuesque wonders, set
landscapes, and panoramic climaxes. Yellowstone is a British Museum of
natural anomalies. The Tetons are composed as The Last Supper. The
Grand Canyon is water's consummate sculpture. Our parks provide essentially a ceremonial experience, through which an informed public passes
properly awed, and exiled from its own feelings.

Park custodians have the same weakness as the rest of us: they love to
name, to isolate, to point out, and to enshrine. A friend and I once hiked
in Capitol Reef toward some formation whose provocative title I have forgotten. Stenciled footprints like a yellow zipper drew us across the
bedrock. The right foot was twice as large as the left, and at first we had a
fine time imagining this lopsided pedant so typical of park planners. The
rest of the slickrock, the junipers, the lichen, the hard earth, and blue sky,

were brilliantly anonymous as ever. But those yellow footprints, marching us toward some scenic aberration, deadened us to the routine loveliness of where we were. The zipper was in poor taste, but it was the failure of our own sensibilities to keep it in scale. Realizing that our goal could only be the climactic letdown, we returned to the parking lot and fled to less ceremonial terrain beyond the park.

It is too bad that things arrowed and tagged lose their vitality. The industrial web is closing in, and never had we more need of the nameless. All American land now lies in some public or private category, but that is not enough. It seems we can only save our last wilderness by culling it from its random occurrence in National Forest and BLM lands, and transferring it to the sharper boundaries and stricter confinements of legislated Wilderness. We can only rescue what's wild by expanding the process of Babel.

It is our twist to an old paradox: that our rage for order craves the very structure our thirst for adventure flees. As technology gathers control, our fences are both tighter and more invisible, our internal map more strung with secret barbed wire. And perhaps the most diehard adventurers, who avoid designated wilderness as fiercely as they avoid parks and monuments, are those in most desperate flight from the plats, grids, and ghostly projections of our own brains.

🦌

"Economic Nature"
Jack Turner

"The conservation movement is, at the very least, an assertion that these interactions between man and land are too important to be left to chance, even that sacred variety of chance known as economic law."

—Aldo Leopold

We live surrounded by scars and loss. Each of us carries around a list of particular offenses against our place: a clear-cut, an overgrazed meadow, a road, a dam. Some we grudgingly accept as necessary, others we judge

mistakes. The mistakes haunt us like demons, the demons spawn aveng-
ing spirits, and the presence of demons and spirits helps make a place our
home. It is not accidental that "home" and "haunt" share deep roots in
Old English, that we speak of the home of an animal as its haunt, or that
"haunt" can mean both a place of regular habitation and a place marked
by the presence of spirits. Like scars, the spirits are reminders—traces by
which the past remains present.

Forty years ago big cutthroats cruised the Gros Ventre River of Jackson
Hole, Wyoming. Now, in late summer, dust blows up the river bed. It's as
dry as an arroyo in Death Valley, a dead river drained by ranchers. Each
autumn much of Jackson Lake, the jewel of Grand Teton National Park, is
a mud flat baking in the sun, its waters drained to irrigate potatoes.
Without good snowfalls each winter the lake could disappear and with
it the big browns, and with those browns, Gerard Manley Hopkins's
"rose moles all in stipple upon trout that swim."[1] The western border of
Yellowstone National Park can be seen from outer space, a straight line cut
through a once fine forest by decades of clearcutting. From the summits
of the Tetons, I see to the west a mosaic of farms scarring the rounded hills
and valleys, as though someone had taken a razor to the face of a beautiful woman. Farther west, the sockeye salmon no longer come home from
the sea. The rivers are wounded by their absence.

These wounds and scars are not random. We attribute the damage to
particular people or corporations or to generalities like industrialization,
technology, and Christianity, but we tend to ignore the specific unity that
made *these* particular wounds possible. This unity lies in the resource
economies of the West: forestry, grazing, mineral extraction, and the vast
hydrological systems that support agriculture. Healing those wounds
requires altering these economies, their theories, practices, and most
deeply and importantly, their descriptions of the world, for at the most
fundamental level the West has been wounded by particular uses of language.

Modern economics began in postfeudal Europe with the social forces
and intellectual traditions we call the Enlightenment. On one level, its
roots are a collection of texts. Men in England, France, and Germany
wrote books; our Founders read the books and in turn wrote letters,
memoranda, legislation, and the Constitution, thus creating a modern
civil order of public and private sectors. Most of the problems facing my
home today stem from that duality: water rights, the private use of public
resources, public access through private lands, the reintroduction of
wolves into Yellowstone National Park, wilderness legislation, the private

cost of grazing permits on public lands, military overflights, nuclear testing, the disposal of toxic waste, county zoning ordinances—the list is long. We are so absorbed by these tensions, and the means to resolve them, that we fail to notice that our maladies share a common thread—the use of the world conceived of as a collection of resources.

Almost everyone agrees the use of public and private resources is out of kilter, but here agreement ends. This absence of agreement is the key to our difficulties, not, for instance, the cost of grazing fees.

A civil society is marked by a barely conscious consensus of beliefs, values, and ideals—of what constitutes legitimate authority, on what symbols are important, on what problems need resolution, and on limits to the permissible. I think of this consensus as a shared vision of the good. Historically, our shared vision of the good derived from shared experience and interests in a shared place. In the West, these "sharings" have vanished—assuming, of course, they ever existed. We share no vision of the good, especially concerning economic practices. One of many reasons for this is the growing realization that our current economic practices are creating an unlivable planet.

The decline in consensus also erodes trust. Trust is like glue—it holds things together. When trust erodes, personal relations, the family, communities, and nations delaminate. To live with this erosion is to experience modernity.[2] The modern heirs of the Enlightenment believe material progress is worth the loss of shared experience, place, community, and trust. Others are less sanguine. But in the absence of alternatives the feeling of dilemma becomes paramount: most of us in the West feel stuck.

Daniel Kemmis's fine book *Community and the Politics of Place* traces some of the West's current dilemmas to the often conflicting visions of Jefferson and Madison, and no doubt some of our dilemmas can be discussed productively in this context. But I think the problems lie deeper. After all, Jefferson and Madison derived their ideas from the works of Enlightenment figures, especially John Locke and Adam Smith, men whose thought was a mixture of classical science, instrumental reason, and Christian revelation.

The heirs of Locke and Smith are the members of the so-called Wise Use movement. Its vigor derives from an accurate assessment: the social order they believe in *requires* Christian revelation, pre-Darwinian science, pre-particle physics, and a model of reason as the maximization of utility. The accuracy of this assessment, in turn, disturbs both liberals and conservatives who wish to preserve Enlightenment ideals while jettisoning the Christian foundations upon which those ideals rest. Unfortunately, that reduces social theory to economics. As John Dunn concluded twenty-five years ago in *The Political Thought of John Locke*, "'Lockean' liberals of the contemporary United States are more intimately than they realize the heirs of the egalitarian promise of Calvinism. If the religious purpose and sanction of the calling were to be removed from Locke's theory, the purpose of individual human life and of social life would both be exhaustively defined by the goal of the maximization of utility" (250). That's where we are now. Instead of a shared vision of the good, we have a collection of property rights and utility calculations.

Since I am a Buddhist, I do not restrict equality to human beings, nor do I justify it by Christian revelation. Nor do I see any reason to restrict "common" (as in "the common good") or "community" to groups of human beings. Other citizens of the West have different understandings and justifications of these key political terms, so part of the solution to the West's differences involves language.

Between Newton and the present, the language of physical theory changed and our conception of reality has changed with it. Unfortunately, the languages of our social, political, and economic theories have endured despite achieving mature formulation before widespread industrialization, the rise of technology, severe overpopulation, the explosion of scientific knowledge, and globalization of economies. These events altered our social life without altering theories *about* our social life. Since a theory is merely a description of the world, a new set of agreements about the West requires some new descriptions of the world and our proper place in it.

Against this background, environmentalism, in the broadest sense, is a new description of the world. The first imaginings of the movement have led to what *Newsweek* has called "the war for the West." Attorney Karen Budd, who often supports Wise Use agendas, says, "The war is about philosophy," and she's right.[3] The fight is over intellectual, not physical, resources. Environmentalists fight to reduce the authority of certain descriptions—e.g., "private property"—and to extend the authority of other descriptions—e.g., "ecosystem." It is the language of pilgrims who entered the wilderness and found not Him, but the Wild.

These new forces have occupied the border of our minds—strange figures claiming high moral ground, like Sioux along the ridges of the Missouri. It's unsettling. Folks employed in traditional economies are circling the wagons of old values and beliefs. Their tone and posture is defensive, as it must be for those who, hurled into the future, adamantly cling to the past.

II

THE PIONEERS WHO SETTLED the West imposed their descriptions on a place they called wilderness and on people they called savages. Neither were, by definition, a source of moral value. The great debates of Jefferson, Madison, Hamilton, and Adams were filled with Enlightenment ethics, revelation, science, political theory, and economic theory. The pioneers brought these ideas west to create a moral and rational order in a new land. Their ideas of what was moral and rational were connected by economics.

The government's great surveys redescribed the western landscape. In 1784 the federal government adopted a system of rectangular surveying first used by the French for their national survey. The result was a mathematical grid: six-mile squares, one-mile squares.[4] Unfold your topo map and there they are, little squares everywhere. Fly over a town or city and you will see people living in a matrix resembling a computer chip. The grid also produced rectangular farms, national parks, counties, Indian reservations, and states, none of which have any relation to the biological order of life.

The grid delighted the pioneers though; they believed a rationalized landscape was a good landscape. It was a physical expression of order and control—the aim of their morality. The idea, of course, was to sell the grid for cash. Indeed, the selling of the grid was the primary reason for its existence. This shifted the locus of the sacred from place to private property. As John Adams said, "Property must be sacred or liberty cannot exist." So the grid was sold to farmers, ranchers, and businessmen, and the places long sacred to the indigenous population simply vanished behind the grid, behind lines arrogantly drawn on paper. With the places gone, the sense of place vanished too—just disappeared.

The sale didn't work out quite as planned. Some land was sold, but often for as little as $1.25 an acre. Other land passed "free" to those who worked it. What was not sold became public land or was reserved to imprison the remnants of the indigenous population. Much of it was simply given to commercial interests.

The railroads alone received 233 million acres. For comparison, consider that Yellowstone National Park's boundaries encompass 2.3 million acres, and that in 1993 our entire national park system—including parks, national monuments, historic sites, historic parks, memorials, military parks, battlefields, cemeteries, recreational areas, lake shores, seashores, parkways, scenic trails, and rivers, in the lower forty-eight *and* Alaska— totals 79 million acres. Consider also that 59 percent of our wilderness areas (which, combined, total 91 million acres) are smaller than Disney World.

Agricultural practices forever destroyed the autonomy of the land sold to farmers and ranchers. Jefferson wrote that "those who labor in the earth are the chosen people of God, if ever He had a chosen people, whose breasts He has made His peculiar deposit for substantial and genuine virtue. It is the focus in which he keeps alive that sacred fire, which otherwise might escape from the face of the earth."[5] God's chosen perceive it good to move water around with irrigation systems; they perceive it good to introduce foreign species of plants and animals; they perceive it good to destroy all that is injurious to their flocks and gardens. In short, they perceive as good that which is good for farmers and ranchers.

Federalists were less convinced of the inherent goodness of farmers, and in retrospect, of course, they were correct. (After all, farmers had burned women at the stake in New England, and, in other parts of the world still boiled and ate their enemies.) Their solution was a federal system of checks and balances. Just as the free market would transform the pursuit of economic self-interest into the common good, so a federal government would transform the pursuit of political self-interest into the common good. Unfortunately, the pursuit of self-interest merely produced more self-interest, an endless spiral that we now recognize as simple greed. In short, the social order of the American West was a mishmash of splendid ideals and pervasive blindness—a rationalized landscape settled by Christians holding private property as sacred and practicing agriculture and commerce under the paternal eye of the federal government. Eventually, of course, these forces proved unequal in power and effect.

Things change. Governmental regulations, commercial greed, and the expanding urban population gobbled up family farms, ranches, and communities, and left in their place industrial agriculture, large tracts of empty land held by banks, subdivisions, and malls. In Wyoming, for instance, only 2 percent to 4 percent of jobs now depend on agriculture.

Things change. The little squares got smaller and smaller as the scale of the social order changed. First there was the section, then the acre, then the hundred-foot lot, then wall-to-wall town houses, then condos. Last year the town of Jackson, Wyoming, contemplated building three-hundred-square-foot housing—about the size of a zoo cage. Most people live in tiny rented squares and the ownership of sacred property is an aging dream. The moral force of private property, derived from owning land, usually large amounts of land, has dropped accordingly. For most people, the problems connected with large holdings of private land are inconsequential. Asking citizens to lament the government's incursion into private-property rights increasingly obliges them to feel sorry for the rich, an obligation that insults their sense of justice.

Things change. The federal system of checks and balances consistently stalls and sabotages federal legislation, making hash of federalism. Every time Congress meets, it is pressured to gut the Clean Air Act and the EPA. Despite widespread regional and national support, twenty years elapsed between the passage of the Endangered Species Act and the reintroduction of wolves in Yellowstone.

Things change. Even the mathematical grid is under attack. The idea that our social units should be defined by mathematical squares projected upon Earth from arbitrary points in space appears increasingly silly. One result is the interest in bioregionalism, the view that drainage, flora, fauna, land forms, and the spirit of a place should influence culture and social structure, define its boundaries, and ensure that evolutionary processes and biological diversity persist.

Things change. A new generation of historians have redescribed our past, deflating the West's myths with rigorous analysis of our imperialism, genocide, exploitation, and abuse; our vast hierarchies of wealth and poverty; the collusion of the rich and the government, especially over water; the biological and ecological ignorance of many farmers, ranchers, and capitalists; and, finally, how our old histories veiled the whole mess with nods to Republican and Jeffersonian ideals. Anyone who bothers to read the works of Donald Worster, Dee Brown, Patricia Nelson Limerick, and Richard White will be stripped forever of the comfortable myths of pioneer and cowboy.[6]

Few, I believe, would deny these changes, and yet in our public discourse of hearings and meetings and newspaper editorials we continue to trade in ideas appropriate to a small homogeneous population of Christian agriculturists occupying large units of land. We continue to believe that politicians represent people, that private property assures liberty, and that agriculture, commerce, and federal balances confer dignity and respect on the West and its people. Since this is largely illusion, it is not surprising that we face problems.

Only one widely shared value remains—money—and this explains our propensity to use business and economics rather than moral debate and legislation to settle our differences. When "the world" shrinks into a rationalized grid stuffed with resources, greed goes pandemic.

Many conservation and preservation groups now disdain moral persuasion, and many have simply given up on government regulation. Instead, they purchase what they can afford or argue that the market should be used to preserve everything from the ozone layer to biodiversity. They offer rewards to ranchers who allow wolves to den on their property, they buy trout streams, they pay blackmail so the rich will not violate undeveloped lands. They defend endangered species and rain forests on economic grounds. Instead of seeing modern economics as the problem, they see it as the solution.

This rejection of persuasion creates a social order wherein economic language (and its extensions in law) exhaustively describes our world and, hence, *becomes* our world. Moral, aesthetic, cultural, and spiritual orders are then merely subjective tastes of no social importance. It is thus no wonder that civility has declined. For me this new economic conservation "ethic" reeks of cynicism—as though having failed to persuade and woo your love, you suddenly switched to cash. The new economic conservationists think they are being rational; I think they treat Mother Nature like a whorehouse.

Ironically, the Enlightenment and civil society were designed to rescue us from such moral vacuums. The Enlightenment taught that human beings need not bow to a force beyond themselves, neither church nor king. Now we are asked to bow to markets and incentives.

Shall we bow to the new king? Can the moral concerns of the West be resolved by economics? Can new incentives for recycling, waste disposal, and more efficient resource use end the environmental crisis? Can market mechanisms restore the quality of public lands? Does victory lie in pollution permits, tax incentives, and new mufflers? Will green capitalism preserve biodiversity? Will money heal the wounds of the West?

One group that answers these questions in the affirmative is New Resource Economics. It welcomes the moral vacuum and fills it with markets and incentives. As economic theory it deserves scrutiny by economists. I am not an economist but a mountaineer and desert rat. Nonetheless, I shall have my say even though the word "economics" makes me hiss like Golem in Tolkien's *The Hobbit:* "I hates it, I hates it, I hates it *forever.*" For I believe classical economic theory, and all the theories it presupposes, is destroying the magic ring of life.

III

IN THE WINTER OF 1992 I flew to Seattle at the generous invitation of the Foundation for Research on Economics and the Environment to attend a conference designed to acquaint environmental writers with the ideas of New Resource Economics. The conference was held amidst a mise-en-scène of assurance and power—tasteful, isolated accommodations, lovely meals, good wine. I felt like a barbarian called to Rome to applaud its splendor.

The best presentations were careful, devastating analyses of the inefficiency and incompetence of the U.S. Forest Service. In sharp contrast were other presentations with vague waves at the preferred vocabulary of self-interest: incentives, market, liberty. They exuded an attitude of "You *see!*" as though the realm of sylvan possibilities was limited to two choices: socialism or New Resource Economics. They were Eric Hoffer's true believers, folks who had seen the light and are frustrated and angry that others fail to see economics as the solution to our environmental plight.

I not only failed to see the light, I failed to understand what was new about New Resource Economics. The theory applies ideas about markets that are now more than two hundred years old. After awhile I had the feeling of watching the morally challenged tinker with notions rapidly disappearing over the horizon of history as they attempted to upgrade one antiquated idea into another. And yet I have little doubt they will succeed. Having just flown over the devastated forests east of Seattle, I wanted to scream, "See the fate of the Earth, the rape of the land!"—but I knew they would respond calmly with talk of incentives and benefits and inefficiency.

Finally I understood. The conference's hidden agenda was to persuade environmental writers to describe nature with an economic vocabulary. They had a theory, and like everyone with a theory, they were attempting to colonize with their theoretical vocabulary, thus eliminating other ways of describing the world.

The conference literature reeked of colonization. Vernon L. Smith's paper, *Economic Principles in the Emergence of Humankind,* describes magic, ritual, and foraging patterns in hunter–gatherer cultures with terms like "opportunity cost," "effort prices," and "accumulated human capital."[7] Michael Rothchild, in *Bionomics: Economy as Ecosystem,* extends economic vocabulary to ecosystems and animal behavior; a niche becomes an organism's "profession," its habitat and food "basic resources," its relations to habitat simply a part of the "economy of nature."

In *Reforming the Forest Service,* Randal O'Toole claims that "although the language used by ecologists differs from that of economists, it frequently translates into identical concepts. Where economists discuss efficiency, decentralization, and incentives, ecologists discuss the maximum power principle, diversity, and feedback loops." O'Toole also maintains that "these very different terms have identical meanings," and he concludes that "ecological systems are really economic systems, and economic systems are really ecological systems" (193).

The redescription of everything with economic language is characteristic of those who sit in the shade of the Chicago school of economics. Thus Richard Posner, in *The Economic Aspects of Law,* colonizes legal issues with economic vocabulary. Regarding children, Posner thinks "the baby shortage and black market are the result of legal restrictions that prevent the market from operating as freely in the sale of babies as of other goods. This suggests as a possible reform simply eliminating the restriction."[8] Bunker, Barnes, and Mosteller's *Costs, Risks, and Benefits of Surgery* does the same for medical treatment.

Indeed, all areas of our social life have been redescribed in economic language. If you like the theory in one area, you will probably like it everywhere. Nor is economic redescription limited to social issues. For example, Robert Nozick, in *The Examined Life,* applies economic language to the question of why we might love our spouse.

> Repeated trading with a fixed partner with special resources might make it rational to develop in yourself specialized assets for trading with that partner (and similarly on the partner's part toward you); and this specialization gives some assurance that you will continue to trade *with that party* (since the invested resources would be worth much less in exchanges with any third party). Moreover, to shape yourself and specialize so as to better fit and trade with that partner, and therefore to do so less well with others, you will want some commitment and guarantee that the party will continue to trade with you, a guarantee that goes beyond the party's own specialization to fit you. (77–78)

In a footnote, Nozick says, "This paragraph was suggested by the mode of economic analysis found in Oliver Williamson, *The Economic Institutions of Capitalism.*"

Why stop with love? In *The New World of Economics* by McKenzie and Tullock, sex becomes a calculated rational exchange.

> [I]t follows that the quantity of sex demanded is an inverse function of price. . . . The reason for this relationship is simply that the rational individual will consume sex up to the point

that the marginal benefits equal the marginal costs. . . . If the price of sex rises relative to other goods, the consumer will "rationally" choose to consume more of the other goods and less sex. (Ice cream, as well as many other goods, can substitute for sex if the relative price requires it.)[9]

So, many men are bores, and what to do? Why bother with arguments, why not just giggle? Unfortunately, too much is at stake.

If we are to preserve a semblance of democracy in the West, we must become crystal clear about how economists colonize with their language.

To start, look at an example of redescription by a theory I disapprove of. Consider, for instance, psycho-babble.

"What did you do today?"

"I cleaned my desk."

"Ah yes, being *anal compulsive* again."

"No, it was just a mess."

"No need to be *defensive*."

"I'm not being *defensive*, I'm just disagreeing with you."

"Yes, but you disagree with me because you have an *unresolved conflict* with your father."

"No, I always got along well with Dad."

"Of course you believe that, but the conflict was *unconscious*."

"There was no conflict!"

"I am not your father! Please don't *cathect* your speech with *projected aggression*."

Ad infinitum. Ad nauseam.

Resource, market, benefits, rational, property, self-interest function the same way as *conflict, unconscious, cathect,* and *projected aggression.* They are simply the terms a particular theory uses to describe the world. By accepting those descriptions, you support and extend the theory. You could decide to ignore the theory, or conclude that the theory is fine in its limited context but shouldn't be extended into others. But if we don't want the fate of our forests decided by bar graphs, we need to cease talking about forests as measurable resources. That does not require you to stop talking to your investment banker about the bar graphs in her analysis of your portfolio.

Economists and scientists have conned us into speaking of trees as "resources," wilderness as a "management unit," and picas gathering grass for the winter because of "incentives." In accepting their descriptions, we allow a set of experts to define our concerns in economic terms and pre-determine the range of possible responses. Often we cannot even raise the issues important to us because the economic language of others excludes our issues from the discussion. To accept this con emasculates not only radical alternatives, but all alternatives. Every vocabulary shapes the world to fit a paradigm. If you don't want nature reduced to economics, then *refuse to use its language.*

This process of theoretical redescription has been termed "colonization" because it privileges one description of the world and excludes others. The Sioux say the Black Hills are "sacred land," but they have found that "sacred land" does not appear in the language of property law. There is no office in which to file a claim for sacred land. If they filed suit, they'd discover that the Supreme Court tends to protect religious belief but not religious practices in a particular place—a very Protestant view of religion.

Language is power. Control people's language and you won't need an army to win the war for the West. There will be nothing to debate. If we are conned into describing the life of the Earth and our home in terms of benefits, resources, self-interest, models, and budgets, then democracy will be dead. What to do? I have five suggestions.

First, refuse to talk that way. It's like smoking, or eating lard. Just say no, and point out that your concerns cannot be expressed in that language. Second, develop a talent for light-hearted humor using economic language. Here again, Thoreau was a prophet. Henry knew a great deal about economics. He read Locke and his followers in both his junior and senior years at Harvard; he was acquainted with the ideas of Smith, Ricardo, Say, and Franklin; and he helped run his family's pencil business when the industry was becoming increasingly competitive and undergoing rapid change. But Thoreau flips economic language on its head. (Remember, the first chapter of *Walden* is titled "Economy.") His "trade" turns out to be with the Celestial Empire; his "enterprises" are inspecting snow storms and sunrises; he "sinks his capital" into hearing the wind; he "keeps his accounts" by writing in his journal; and he gleefully carries the cost of rye meal out to four decimal places: $1.0475. Nothing is fixed, all is metaphor, even economics.

Third, become so intimate with the process of economic description, you *experience* what's wrong with it. Since economics is a world of resources—physical resources, cultural resources, recreational resources, visual resources, human resources—our wonderfully diverse, joyful world must be reduced to measurable resources. This involves abstraction, translation, and a value. Just as time is abstracted from experience and rendered mechanical (the clock) so it can be measured, space is abstracted from place and becomes property: measurable land. In the same way, trees are abstracted into board-feet, wild rivers are abstracted into acre-feet, and beauty is abstracted into a scene whose value is measured by polls. Economics reduces everything to a unit of measurement because it requires that everything be commensurate—"capable of being measured by a common standard"—its standard. The variety of these calculable units may be great—board-feet, time, tons, hours—but all of these units can be translated into a common value similar to the way different languages can be translated. Both types of translation require something common. In linguistic translation, it is meaning; in economic translations, it is money—not the change in your pocket, but the stuff that blips on computer screens and bounces off satellite dishes from Germany to Japan in less than a second. An hour's labor is worth a certain amount of money; so is three hundred board-feet of redwood.

Once everything is abstracted into commensurate units and common value, economic theory is useful. If the value of one kind of unit (computer chips) grows in value faster than another kind of unit (board-feet), economic theory says translate board-feet into money into computer chips. In ordinary English: Clear-cut the last redwoods for cash and buy Intel stock. If you don't like deciding the fate of redwoods by weighing the future of Intel, then you probably won't like economics.

Refuse these three moves—the abstraction of things into resources, their commensurability in translatable units, and the choice of money as the value of the units—and economic theory is useless.

Once you understand the process, it's easy to recognize examples. For instance, in *Reforming the Forest Service,* Randal O'Toole describes walking in the mountains as a wilderness experience using a recreational resource that generates benefits: cash and jobs (206). These benefits are compared to other possible uses of the resource, say, grazing and logging, that generate other benefits. The benefits can then be compared. This provides a rational basis for budget maximization. Your walk in the Tetons becomes, by redescription, an economic event.

A fourth way to subvert economic language is to realize that nothing of great value is either abstract or commensurate. Start with your hand. The workman's compensation office can tell you the value of your hand in dollars. Consider your daughter. An insurance company or litigation lawyer can tell you her value in dollars. What is your home place worth? Your lover's hair? A stream? A species? Wolves in Yellowstone? Carefully imagine each beloved person, place, animal, or thing redescribed in economic language. Then apply cost-benefit analysis. What results is a feeling of sickness familiar from any forest sale or predator-control proposal. It is the sickness of being forced to use a language that ignores what matters in your heart.

Finally, realize that describing life—the completely individual, unique here-now alive *this*—with abstractions is especially dissonant. Consider the "resources" used in a biology class. The founder of experimental physiology, Claude Bernard, said that the man of science "no longer hears the cry of animals, he no longer sees the blood that flows, he sees only his idea and perceives only organisms concealing problems which he intends to solve." He sees only the idea that will give him something to do in the world. Meanwhile the screams of animals in laboratory experiments are redescribed as "high-pitched vocalizations."

In an extraordinary essay, "Pictures at a Scientific Exhibition," William Jordon, an entomologist, describes his graduate education and the ghastly (his word) treatment of animals it required.

> Fifteen years ago I saw several of my peers close down their laboratory for the evening, and as they cleaned up after the day's experimentation they found that three or four mice were left over. The next experiments were not scheduled for several weeks, and *it wasn't worth the cost and effort to keep the mice alive until then*. My friends simply threw the extras into a blender, ground them up, and washed them down the sink. This was called the Bloody Mary solution. Several days ago I talked with another old peer from my university days, and she informs me that the new, humane method for discarding extra mice in her lab is to seal them in a plastic bag and put it in the freezer.
>
> I repeat: the attitude toward nonhuman life has not changed among experimental biologists. Attitude is merely a projection of one's values, and their values have not changed; they do not respect life that is not human. (199, my emphasis)

Science, including economics, tends to reduce nonhuman life to trash. The screaming animals, the dead coyotes, the Bloody Mary mice, the stumps, the dead rivers—all are connected by these processes of abstraction, commensurability, and financial value. There is no logical necessity for us to describe the world this way. The Apaches didn't do it, and we need to reach a point where we don't do it either.

We need to find another way of describing the world and our experience in it. Leave this pernicious, mean-spirited way of talking behind. One of my heroes said he could imagine no finer life than to arise each morning and walk all day toward an unknown goal forever. Basho said this *is* our life. So go for a walk and clear the mind of this junk. Climb right up a ridge, over the talus and through the whitebark pine, through all those charming little grouse wortleberries, and right on into the blue sky of Gary Snyder's *Mountains and Rivers Without End:*

> *the blue sky*
> *the blue sky*
>
> *The Blue Sky*
> is the land of
> OLD MAN MEDICINE BUDDHA
> where the eagle
> that flies out of sight,
> *flies.*[11]

IV

TRAVELING TO THAT CONFERENCE last winter, I found the approach to Seattle from the east to be infinitely sad. Looking down at those once beautiful mountains and forests, so shaved and mowed down they look like sores, I didn't care if the land below was public or private, if the desecration was efficient or inefficient, cost beneficial, or subsidized, whether the lumber products were sent to Japan or used to build homes in Seattle. I was no longer interested in that way of looking at the world. Increasingly, I am a barbarian in the original sense of the Greek word— one who has trouble with the language of civilization. So, slowly and reluctantly, I am burning bridges to the past, all the while noticing, as if in penance, that the ideas and abilities of a trained pedant follow close as shadows.

A passage from an obscure journal by the philosopher Nelson Goodman often occupies my mind. "For me, there is no way which is the way the world is; and so of course no description can capture it. But there are many ways the world is, and every true description captures one of them."[12]

The universe we can know is a universe of descriptions. If we find we live in a moral vacuum, and if we believe this is due in part to economic language, then we are obligated to create alternatives to economic language. Old ways of seeing do not change because of evidence; they change because a new language captures the imagination. The progressive branches of environmentalism—defined by an implacable insistence on biodiversity, wilderness, and the replacement of our current social grid with bioregions—have been sloughing off old ideas and creating one of many possible new languages.

Emerson started the tradition by dumping his Unitarian vocabulary and writing "Nature" in language that restored nature's sacredness. Thoreau altered that vocabulary further and captured our imagination. The process continues with the labor of poets, deep ecologists, and naturalists. It is not limited to radical environmentalism, however; it includes many who are only partially sympathetic to the radical cause. Michael Pollan, for example, tells us in *Second Nature* that science has proposed some new descriptions of trees as the lungs of the Earth. And radical economist Thomas Michael Power suggests in *The Economic Pursuit of Quality* that "economy" might be extended beyond commerce. The process is enforced when Charles F. Wilkinson, in *The Eagle Bird,* suggests changes in the language of law that would honor our surrender to the beauty of the world and of emotion.

Imagine extending the common in "common good" to what is common to all life—the air, the atmosphere, the water, the processes of evolution and diversity, the commonality of all organisms in their common heritage. Imagine extending "community" to include all the life forms of the place that is your home. Imagine "accounting" in its original sense: *to be accountable.* What does it mean to be accountable, and to whom and to what purpose? What's "a good deal" with the Universe? Imagine an economics of need. Instead of asking "What is this worth?" ask "What does this forest need?" "What does this river need?"

Consider Lewis Hyde's beautiful description of an Amish quilt sale: "A length of rope stretched around the farm yard full of household goods. A little sign explained that it was a private auction, in which only members of the Amish community were allowed to bid. Though goods changed hands, none left the community. And none could be inflated in value. If sold on the open market, an old Amish quilt might be too valuable for a young Amish couple to sleep under, but inside that simple fence it would always hold its value on a winter night."[13]

"Hold its value on a winter night"? What's happening here?

It's as simple as that rope and a group of people deciding to place aspects of their shared experience above economic values determined by the open market. They don't ignore economic value—there is still a price, bidding, and competition—but it is restrained by a consensus of appreciation a wider market would ignore.

Although this example comes from a religious community, its power does not turn on religion; although it comes from an agricultural community, it does not turn on agriculture. It turns on two things: shared experience and shared place—the politics of locale. As does the Bill of Rights, the rope creates a limit with standards and values shared by the community. We need to imagine an immense fugue of variations on that simple fence, each creating a new world.

These imaginings will be the worthy labor of poets and thinkers and artists whose primary task, it seems to me, is to extend those qualities we value most deeply—the source of our moralities and spiritual practices—into what we call "the world." Many will find that source is empty, drained like the great aquifers that water our greed. Others will discover links between their integrity and that of an ecosystem, between their dignity and the dignity of a tree, between their desire for autonomy and the autonomy all beings desire, between their passions and the wild processes that sustain all life.

Extend these moral and spiritual sources into nature and the spirits of each treasured place will *speak* as they have always spoken—through art, myth, dreams, dance, literature, poetry, craft. Open the door and they will transform your mind instantly. If children were raised hearing stories of spotted owls, honoring them with dances, imagining them in dreams, and seeking the power of their gaze, then spotted owls would speak to us, transformed by mind into *Our-Form-of-Life-At-The-Place-of-Spotted-Owls*.

Then we wouldn't have to worry about clear-cutting spotted owl habitat. And when wildfires articulated their needs, we would not drown them in chemicals. When wild rivers spoke, they would be cleared of dams, and the salmon would come home from the sea.

Dig in someplace—like a great fir driving roots deep into a rocky ridge to weather storms that are inseparable from the shape of its roots. Allow the spirits of your chosen place to speak through you. Say their names. Say Moose Ponds, Teewinot, Pingora, Gros Ventre, Stewart Draw, Lost River. Speak of individuals—the pine marten that lives in the dumpster, the *draba* on the south ridge of the Grand Teton. Force the spirits of your place to be heard. Be hopeful. Language changes and imagination is on our side. Perhaps in a thousand years our most sacred objects will be illuminated floras, vast taxonomies of insects, and a repertoire of songs we shall sing to whales.

It is April and still cool beside Deer Creek in the Escalante country. Around me lies last year's growth, old sedges and grasses in lovely shades of umber and sienna. Beside me stands an ancient Fremont cottonwood. At the tips of its most extended and fragile branches, bright against a cobalt sky, are the crisp green buds of spring.

NOTES

1. From Gerard Manley Hopkins's poem "Pied Beauty."
2. See Anthony Giddens, *The Consequences of Modernity.* Trust doesn't disappear, but its focus moves from kin and community to abstract systems, especially money and a culture of experts.
3. Bill Turque, "The War for the West," *Newsweek* 118, no. 14 (September 30, 1991): 18; Florence Williams, "Sagebrush Rebellion 2."
4. Paul Shepard describes this process in "Varieties of Nature Hating," chapter 7 of *Man in the Landscape.*
5. Quoted in Kemmis in *Community and the Politics of Place,* 20.
6. See Worster, *Under Western Skies;* Brown, *Bury My Heart at Wounded Knee;* Limerick, *The Legacy of Conquest;* and White, *It's Your Misfortune and None of My Own.*
7. Opening address at the 1991 Mont Pelerin Society meeting, Big Sky, Montana.
8. Quoted in Kenneth Lux's *Adam Smith's Mistake,* 202.
9. Quoted in Lux, 203.
10. Quoted in Evernden, *The Natural Alien,* 16.
11. Snyder, *No Nature,* 80.
12. Goodman, "The Way the World Is," 56.
13. Hyde, "Laying Waste to the Future."

From
Work in Progress
David Brower

A mile of mountain wall spills out of the Wyoming sky beyond a wide meadow, a meadow edged with wonder this morning when a small boy's excited cry *moose!* woke us and we watched mother and calf leisurely browse their way downstream, ford, and then disappear into a tangle of cottonwood, aspen, fireweed, and lodgepole. They were not exactly a graceful pair, for nature had something else in mind than mere grace of line when the moose was designed. But they graced the place where we saw them and added to it a new dimension of wildness and of space. A moose needs a lot of wild space and here she found it, in a place that is just about as much the way it was when trappers first saw it as a place could be and still be part of a national park a million people see each year.

It was three years ago that the boy saw his first moose here. Now his age had doubled without a moose having recrossed his ken; yet he knew exactly, without hesitating a moment, what the cow and calf were and with no rack of antlers to guide him. The image fixed well, as wild images do, on that perfectly sensitized, but almost totally unexposed film of his mind. The same thing would happen to any other small boy, given the chance, and the composite image of a thousand such experiences would enrich his living in the civilized world so thinly separated from the wildness the boy was designed to live with.

But where will the chance to know wildness be, when this boy is himself a father, when a generation from now he is seeking out a place in which to expose his own six-year-old to wonder? How much of the magic of this, the American earth, will have been dozed and paved into oblivion by the great feats of engineering that seem to come so much more readily to hand than the knack of saving something for what it is?

Man's marks are still few here, but they are being made faster and faster. The cabin hewed with patient care has mellowed, and the road to it has not burgeoned beyond the two tracks that led there when it was new. The stream has claimed the bridge that once crossed it; twenty-year-old pines grow on one of the approaches, and beavers have built and used and abandoned a lodge on the other. The power line is hardly more permanent than the fence that fell and now moulders in the meadow. The highway is so far away that the drone of cars can hardly be heard above the stream music. Silence closes in soon after the sight-seeing planes pass by the front of the great range.

But each year the silences are briefer. The throng that comes, grows larger, needs more, and the forest and meadow make way to accommodate them. Wider highways speed people through faster and crowd out the places where the cow has dropped her calf for all the generations since the ice retreated, and where the trumpeter swan could inform her cygnets of those few things the evolutionary force had not already told them. Here where the blue vault arches over the wildest and least limited open space and beauty, even here man's numbers are taming and limiting with greater and greater speed, heedless of the little losses that add up to deprivation.

Again and again, the challenge to explore has been met, handled, and relished by one generation—and precluded to any other. Although Thomas Jefferson argued that no one generation has a right to encroach upon another generation's freedom, the future's right to know the freedom of wilderness is going fast. And it need not go at all. A tragic loss could be prevented if only there could be a broader understanding of this: that the resources of the earth do not exist just to be spent for the comfort, pleasure, or convenience of the generation or two who first learn how to spend them; that some of the resources exist for saving, and what diminishes them, diminishes all mankind; that one of these is wilderness, wherein the flow of life, in its myriad forms, has gone on since the beginning of life, essentially uninterrupted by man and his technology; that this, wilderness, is worth saving for what it can mean to itself as part of the conservation ethic; that the saving is imperative to civilization and all mankind, whether or not all men yet know it.

From
The End of Nature
Bill McKibben

We can no longer imagine that we are part of something larger than ourselves—that is what all this boils down to. We used to be. When we were only a few hundred million. or only a billion or two, and the atmosphere had the composition it would have had with or without us, then even Darwin's revelations could in the end only strengthen our sense of belonging to creation, and our wonder at the magnificence and abundance of that creation. And there was the possibility that something larger than us—Francis's God, Thoreau's Benefactor and Intelligence, Peattie's Supreme Command—reigned over us. We were as bears—we slept less, made better tools, took longer to rear our young, but we lived in a world that we found made for us, by God, or by physics and chemistry and biology, just as bears live in a world they find waiting for them. But now we make that world, affect its every operation (except a few—the alteration of day and night, the spin and wobble and path of the planet, the most elementary geologic and tectonic processes).

As a result, there is no one by our side. Bears are now a distinctly different order of being, creatures in our zoo, and they have to hope we can figure out a way for them to survive on our hot new planet. By domesticating the earth, even though we've done it badly, we've domesticated all that live on it. Bears hold more or less the same place now as golden retrievers. And there is nobody above us. God, who may or may not be acting in many other ways, is not controlling the earth. When He asks, as He does in Job, "Who shut in the sea with doors . . . and prescribed bounds for it?" and "Who can tilt the waterskins of the heavens?" we can now answer that it is us. Our actions will determine the level of the sea, and change the course and destination of every drop of precipitation. This is, I suppose, the victory we have been pointing to at least since the eviction from Eden—the domination some have always dreamed of. But it is the story of King Midas writ large—the power looks nothing like what we thought it would. It is a brutish, cloddish power, not a creative one. We sit astride the world like some military dictator, some smelly Papa Doc—we are able to wreak violence with great efficiency and to destroy all that is good and worthwhile, but not to exercise power to any real end. And, ultimately, that violence threatens us. Forget the interplanetary Rose Bowl; "man's synthetic future" has more to do with not going out in the sun for fear of cancer.

But the cancer and the rising sea level and the other physical effects are still in the future. For now, let's concentrate on what it feels like to live on a planet where nature is no longer nature. What is the sadness about?

In the first place, merely the knowledge that we screwed up. It may have been an inevitable divorce: man, so powerful, may not have been meant to live forever within the constraints of nature. It may have been an inevitable progression—man growing up to be stronger than his mother, nature. But even inevitable passages such as these are attended by grief. Ambition, growth take us away from old comforts and assurances. We are used to the idea that something larger than we are and not of our own making surrounds us, that there is a world of man and a world of nature. And we cling to that idea in part because it makes that world of men easier to deal with. E.B. White, in one of his last essays, written from his saltwater farm near Mt. Desert in Maine, said that "with so much disturbing our lives and clouding our future . . . it is hard to foretell what is going to happen." But, he continued, "I know one thing that *has* happened: the willow by the brook has slipped into her yellow dress, lending, along with the faded pink of the snow fence, a spot of color to the vast gray-and-white world. I know, too, that on some not too distant night, somewhere in pond or ditch or low place, a frog will awake, raise his voice in praise, and be joined by others. I will feel a whole lot better when I hear the frogs." There may still be frogs—there may be more frogs, for all I know—but they will be messengers not from another world, whose permanence and routine can comfort us, but from a world that is of our own making, as surely as Manhattan is of our own making. And while Manhattan has many virtues, I have never heard anyone say that its sounds make you feel certain that the world, and you in it, are safe.

Anyway, I don't think that this separation was an inevitable divorce, the genetically programmed growth of a child. I think it was a mistake, and that consciously or unconsciously many of us realize it was a mistake, and that this adds to the sadness. Many have fought to keep this day from coming to pass—fought local battles, it is true, perhaps without realizing exactly what was at stake, but still understanding that the independent world of nature was gravely threatened. By the late 1960s an "environmental consciousness" had emerged, and in the 1970s and 1980s real progress was being made: air pollution in many cities had been reduced, and wilderness set aside, and Erie, the dead lake, that symbol of ultimate degradation, rescued from the grave.

So there is the sadness of losing something we've begun to fight for, and the added sadness, or shame, of realizing how much more we could have done—a sadness that shades into self-loathing. We, all of us in the First World, have participated in something of a binge, a half century of unbelievable prosperity and ease. We may have had some intuition that it *was* a binge and the earth couldn't support it, but aside from the easy things (biodegradable detergent, slightly smaller cars) we didn't do much. We didn't turn our lives around to prevent it. Our sadness is almost an aesthetic response—appropriate because we have marred a great, mad, profligate work of art, taken a hammer to the most perfectly proportioned of sculptures.

🦌

"The Forest of Forgetting"
Guy Hand

When I first set eyes on the Highlands of Scotland only the sky seemed alive, animated by the brooding advance of storms. Through the mist I could find nothing else to focus on, not a house or a fence or a tree—just the rise and fall of vacant land. Rain collected in rivulets. It tumbled like tears from bare stone. At that moment I would have sworn it was the saddest place on earth.

I hadn't realized I'd stumbled into a kind of Highland fable, into a treeless land with a forest story. I could feel it in the wind as it blew over the thin, rock-strewn soil; smell it in the smoke of coal fires; and hear it in the wail of Highland pipes. Yet I couldn't make sense of the meaning until I saw it as a look in my Scottish wife's eyes.

For Mairi the Highlands are home. Her black hair, soft brogue, and dark eyes betray a long ancestry there, of the Picts, the ancient people who stopped the Roman armies' northern advance at Hadrian's Wall. Mairi still carries that fight in her eyes, and a fierce love for that land. We met there, and fell in love. She helped me to see her homeland through a native's eyes and when she gave that homeland up for mine, I had hoped to do the same. But then, there were the trees.

The instant we climbed out of Idaho sagebrush and into a dense stand of pine, in the Sawtooth Wilderness Area, I knew something was wrong. Mairi fell silent. Her pace slowed. I glanced over my shoulder to find the distance between us filled with shadow and half-light. She had hunched her shoulders and dropped her head. She moved with the wary posture of stalked prey. As she passed through a saber of light I could clearly see the fear in her eyes. I waited for her, but she walked passed, pointed to a clearing, and by way of explanation, whispered "too many trees." Neither of us had known, until that moment, that Mairi held a secret dread of wooded land.

I felt as if I'd failed her, unable to convey the closed-in sense of sanctuary I'd always felt in that forest, the way, even as a child, the thick mat of pine needles and jigsaw bits of bark felt luxurious under my feet; the way the trees provided shelter against wind and mid-day glare; the way sounds were both softened and clarified; the way air held the sweet scent of pitch and the flutter of wings.

On the scree and boulder slopes above tree line, the tension drained from her face. She looked off into a landscape she could again understand: open country, treeless country, country filled with nothing more than grass, rock, and sky. It was only later, after peering more deeply into her Highland past, that we learned forests were part of her history too, a forest lost to centuries of forgetting.

At first glance the Highlands and the Sawtooths have little in common below treeline, yet that landscape of heathered moors, shimmering lochs, and bare-rock mountains comes as close as it gets to a British version of wilderness. There are fewer people there and the land has suffered less recent development than anywhere in Europe. After a storm, when the air is crystalline, the sky Atlantic blue, and remnant clouds catch in the folds of far away peaks, the Highlands appear as primal, as unsullied as the granite ridges that loom over the little Sawtooth town of Stanley. Yet, the Highlands are nothing of the kind.

When I stared across that vast, vacant land, I assumed that the truth of the place, the bedrock of Scottish history, lay exposed before me. I was wrong. A thick wilderness of trees once covered Highland bedrock, a forest as grand as any on earth. Elm, ash, alder, and oak shaded the low-lying coastal plains and inland valleys; aspen, hazel, birch, rowan, and willow covered the hills; and beautiful, red-barked Scots pine clung to the glacial moraines and steep granite slopes. The Romans called it the Forest of Caledonia, "the woods on heights," and it clung to Scottish soil for millennia.

People began weaving their lives among the trees soon after the forest claimed the land from Ice Age glaciers. Settlers built homes from oak and hazel, and furniture from elm and alder. They wove baskets from willow and built boats from pine. The Gaelic-speaking Celts fashioned an alphabet connected, for mnemonic reasons, to the names of trees, and used it to write of the reverence they felt for them:

Birch, smooth, blessed,
proud, melodious,
lovely is each entangled branch
at the top of your crest

Like most indigenous cultures, theirs developed through a long and close connection to land. The early Scots saw the lives of trees interlocked with their own. Whether innate or hard-won, they perfected a balanced, reciprocal relationship with the forest, and took from it knowing their own health depended on its preservation. Highland historian James Hunter believes their environmental awareness was unique, predating any other in Europe by hundreds of years.

With the coming of the English and the Industrial Revolution everything changed. Sixteenth-century England was hungry for wood. Empire building had depleted their forests, and as English woodsmen worked their way north, into the Highlands, they brought with them, not only axes, but a profoundly different philosophy of nature—a view aggressively and breathtakingly anthropomorphic, a view that pictured everything on earth as intended for "the benefit and pleasure of man," and untamed woodland as something to be feared, exploited, and, if necessary, erased. Literature of the time bristled with references to "degenerated nature," the "deformed chaos" of woodland, and odes to trees far different from those of the Celts:

> . . . haughty trees, that sour
> The shaded grass, that weaken thorn-set mounds
> And harbour villain crows . . .

The English saw, in the Highlands, not only land darkened with trees, but incivility. They called the native Highlanders "savages," (from the Latin root silva, meaning forest) and their trees "an excrescence of the earth, provided by God for the payment of debts." Through the axe, the Highlands and its people were to be cleansed of chaos and shown the path to culture.

The English, of course, were not the originators or sole practitioners of the belief that culture must be built only on the stumps of trees (a notion as old as Plato). Theirs was neither an inherent or exclusive malice: many Lowland Scots and Highland chiefs soon found profit in the felling of woodland. Nor were they the first to fell trees in the Forest of Caledonia: Over the centuries, the forest had been reduced in size slowly, by a cooling climate and the axes of both settlers and prior invaders—but these English axes were swung with the force of moral imperative and took a far greater toll on the native landscape than any before.

Vast areas were stripped of wood. In Scotland and Ireland both, the land was, in places, so altered that maps had to be redrawn, and as they were, Gaelic place names, which had been used to describe the land directly—the curve of a hill, a stand of birch, or the presence of a stream—were replaced with Anglicized names that had little or no relationship to the terrain they labeled. Eventually, as English influence became pervasive, the Gaelic language itself was suppressed. As forests disappeared, so did the very words used to describe them.

The Highlands were devastated. Stone houses and coal fires replaced those of wood. Soils, exposed to harsh winds and rain, washed into streams and rivers, leaching fertility, destroying fisheries. Erosion cut, in many places, to bedrock. Woodland species—bear, reindeer, elk, moose, beaver, wild boar, wild ox, wolf (the last killed in 1743), crane, bittern, great auk, goshawk, kite, and sea-eagle—vanished. And as the land became impoverished, so did the human population. Many left; others were forcibly removed in what is called The Highland Clearances, as landlords realized sheep would be more profitable than tenants on the now bankrupt Highland soil. As the trees disappeared so did the culture dependent on it. Poet Norman MacCaig wrote:

> Greenshank, adder, wildcat, guillemot, seatrout,
> fox and falcon—the list winds through
> all the crooks and crannies of this landscape, all
> the subtleties and shifts of its waters and
> the prevarications of its air—
> while roofs fall in, walls crumble, gables
> die last of all, and man becomes,
> in this most beautiful corner of the land,
> one of the rare animals.

By 1773, when Dr. Samuel Johnson toured the Highlands with James Boswell, the landscape was, in Johnson's words, a "wide extent of hopeless sterility." He remarked that one was as likely to see trees in Scotland as horses in Venice. By the beginning of the twentieth century Britain supported the lowest percentage of woodland in all of Europe, a mere 4%, and today, having been stripped of more than 99% of its original forest, Mairi's Highland home is one of the most deforested lands in the world.

On the Hebridian island of Barra, west of the mainland, Mairi once led me through a mile of heather to a comfortably rounded block of stone she often sat on as a child. Gone from the island some fifteen years, she negotiated that sea of sameness without hesitation. She bore down on her rock with the steadfast assurance of a ship to home port, and once reunited, traced a patch of yellow and green lichen that clung to her rock's surface and told me how she had sat in that spot, immersed in solitude and forever views for whole afternoons. Mairi's eyes grew moist, and again they reminded me of how deeply the Scottish landscape had imprinted itself on her. Nine years away from Scotland, she still clings to the memory of that land as tightly as lichen.

Yet if there is such a thing as an ancestral memory of aboriginal land, Mairi's had been lost, cut away with the trees. At first I saw her reaction to Idaho forest as a surprising, yet simple response to unfamiliar ground. Now I've come to see it as a kind of cultural forgetting, a way she and her clan severed themselves from a too-painful past—as a wolf chews through its own leg to escape a trap—way of making the present bearable, a way of devaluing loss and finding solace in what is left. In my Idaho forest she may have needed to see menace, to see claws and talons in the guise of pine boughs, to pad the reality of the land she loves. I've since found many Scots with the same unsettled attitude toward trees.

The twentieth century push for national parks in Scotland has met with determined resistance and failure. As British writer Richard Mabey put it, "The fear of 'letting nature go' reaches very deep in Britain. When, a few years ago, the idea of 'set-aside' was first mooted as a solution to the problems of surplus arable land, all kinds of bizarre fears began to surface about the British countryside being 'swallowed up.' . . . I sometimes wonder how those that perpetuate this myth imagine that Britain cloaked itself with woodland 10,000 years ago, long before the invention of foresters."

Foresters? As I stood on that first Highland hillock, I could never have imagined forest, let alone foresters. But I hadn't seen all there was to see. As I searched for something in that landscape to cling to, I eventually found what I assumed were scraps of native woodland, clusters of trees growing in odd places, in odd, geometric shapes, like patches on worn tweed. At the same time, I noticed the contentious attitude locals had toward them. I heard those trees discussed in pubs derisively: they blocked views, hindered grazing, harbored vermin, and were "bloody unnatural."

I, on the other hand, was instantly drawn to those ragged clusters of trees, those small islands of familiarity in an otherwise strange ocean of gorse, heather, and bracken. I soon found they weren't natural woodland at all, but "afforestation" projects administered by a governmental organization, the Forestry Commission. Most were inaccessible, surrounded by fence or flung deep into private land, but on the Isle of Skye, near the town of Portree, I found a patch open to the public. A small sign pointed to a "Forest Walk."

I followed the sign from the road toward a wall of trees. At my feet delicate ferns had displaced coarse heather. The wind, which seemed a ceaseless feature of the Highlands, dropped to a whisper. I couldn't help but smile. Yet as I walked I began to pick out a symmetry that had eluded me from the road: The trunks of those first trees stood exactly five feet apart, and as I passed they aligned themselves with the rest of the forest in straight rows, like Iowa corn. When I crossed the threshold into the woods themselves, through that first facade of green, all hope of spending time in familiar terrain vanished. This forest was a mirage. The green was gone, replaced with a deadened monochrome of brown, the earth as bare and lifeless as a cellar floor. Identical tree trunks rose with the regularity of concrete pillars, bristling with dead branches. Not a speck of life shown to the height of thirty feet, and only there, as the trees met sunlight, did they sprout green needles. The effect was chilling. A dead heart cloaked in green. An illusion. The great Caledonian Forest in reverse.

The Forestry Commission was established in 1919 with one purpose, to grow trees as fast and efficiently as possible. During the First World War Britain faced an acute lack of timber (turning many tattered remains of the forest of Caledonia into ammunition boxes), and shortly thereafter began an ambitious reforestation program. The program is in place today. Land is plowed and planted with exotic conifers—mostly Sitka spruce from British Columbia. Helicopters often fertilize the trees, which are harvested on a thirty-five-year rotation by machines that both cut and stack them. Most end as pulp, paper, or fiber board, their quality far too low for other uses. Although a recent push to diversify the forests and introduce more environmentally sound practices has begun, two and a half million acres of Scotland are now covered in a monoculture of plantation trees.

I didn't fully comprehend the brutishness of this method of British industrial forestry until, a few miles from my forest walk, I found earth recently prepped and planted with young trees. Planting plows had cut four-foot-deep troughs into the black peat soil. Ankle-high spruce trees stood among the heavy lumps of peat in the rows between troughs. I struggled to the spine of a ridge, jumping rain-filled moats every few feet, and peered out on hundreds of them, hundreds of parallel ditches that cut straight downhill and then back up the next slope and on to the horizon. Virtually all the land I could see was shredded, as if by giant claws, the hills flayed like the haunches of a ravaged animal.

I explored several other forest plantations, but found all of them to be equally hideous. Many older ones were impenetrable, trees planted so close together nothing larger than a vole could enter. Russian beetles had infested others, causing them to lose even their thin veneer of green. They stood as straight and dead as a forest of last year's Christmas trees. All the plantations were sad, sickly substitutes for true forest. But that was *my* reaction, the reaction of someone with a memory of intact forest, someone lucky enough to have grown up near the largest wilderness areas in America. The average Scot might never see forest more real than this, may never get beyond the assumption that once-you've-seen-one-tree-you've-seen-them-all, may never understand that a true forest holds wonder as well as wood.

Along the banks of the Boise River, near where I grew up, a dense corridor of black cottonwood forest once stood. Captain Benjamin Bonneville and a group of French explorers spotted it in 1883. Having traveled an arid ocean of sage, Bonneville climbed a ridge, saw that wide, meandering swath of green, and called out *"Les bois! Les bois! Voyez les bois!,"* "The trees! The trees! Look at the trees!"

I have no memory of that forest. By the time I was born it was in tatters, broken, and all but gone. I *do* remember, rooted along the banks of that river, overgrazed, weedy lots, filled not with trees, but cow shit and rusting car bodies, the breeze weighted with the sour stench of rendering plants and open sewers. Mine are *living* memories, memories thick with the sights, sounds, and smells of direct contact to that place. They are also my *only* memories, and though grade school history lessons taught me that the river and this town were named for that missing forest, my memories superseded them, and carried the intimation that riverbanks, *all riverbanks,* by their nature, bred worthless ground.

Once primeval forests are cut, the land loses its ability to hold wonder, and as that wonder bleeds away so does human caring. I could see it draining from the banks of the Boise River and all but gone in the flayed, ruptured landscape of the Scottish Highlands.

Both are places whose forests are fading from memory, and ultimately it may be the forgetting, not the loss of forests, that breaks the connection between the wild world and ourselves. Once the look of mature woodland, the sweet resinous smells, the chirps and squeals, the sense of sanctuary fall from memory, we have no instrument of comparison, no place that we can confidently call uncorrupted, nowhere that hasn't been impoverished by our own limited grasp of the way the world works. And once forgotten, such wonders may be impossible to imagine back into being.

Boise has tried to imagine its riparian forest back, and although the "Greenbelt" riverbank restoration project has received much national praise, it has veered from a return to the original forest to that of an urban park, sprouting short-sighted commercial and residential floodplain developments where trees should be.

The Highlands of Scotland have seen attempts at restoration as well, yet a much longer time has elapsed since the forest fell. For most the Highlands are still a mistaken wilderness disguised by forgetting, a vast, devastated land, made beautiful by myth and melancholy light.

When I first gazed into that light, I never dreamed I was looking at the past and possible future of American forests. Hidden in the shadows was a forest story not yet finished. America is heir to that story, and the ragged woodland along the Boise River, the Cove Mallard timber fights in Central Idaho, the clear cuts on the Olympic Peninsula, the fallen redwoods of Northern California, and all the other American forest stories are chapters written in the same script. Our sylvan attitudes germinated in British forests, and in the barren Highlands you can track their evolution and see, in harsh terms, their ultimate result.

You might think I'm overreacting here, but first consider that the largest ponderosa pine in the world stands, not in a wilderness area, but sliced and cradled in a Boise park. Two ten-foot sections of that tree support a beamed roof under which the "butt cut flare" (the widest section of the tree) is displayed on its side, to show nearly four centuries of tree rings. A bronzed plaque above the tree declares this "THE WORLD'S LARGEST PONDEROSA" and two other plaques, nailed to each of the flanking upright sections, record statistics. One plaque explains that the tree, when cut, was 9 feet 4 inches wide, 126 feet high, and 376 years old. The other plaque lists important dates in human history, and the tree's rings are marked with numbers corresponding to each date: The 1620 landing of the pilgrims at Plymouth Rock, marked at the center of the tree; the 1644 English renaming of New York, four inches out; the 1776 signing of the Declaration of Independence, another two and a half feet from the center; the 1805 entering of Lewis and Clark into Idaho, a few more inches; and so on to the tree's outer bark.

This cross section seems, at first, an odd symbol for remembering (irony clings to any tree felled, sliced, and displayed to show proof of its longevity) and although it shows, in the exquisite precision of tree rings, that the lives of trees and the lives of humans have shared much history, and that trees have bore witness to the coming of European civilization to

America, few forests have endured the results of that landing. When the first English colonists set foot in Massachusetts, they brought the British forest story to American shores. The vocabulary used to describe the American landscape echoed that of the Highlands, the Plymouth Colony calling the land they found a "hideous and desolate wilderness . . . full of woods and thickets." The problems faced "betwixt the American colonists and the aborigines of that country" were not very different from those faced with the Highlanders, nor was the rationale for occupying their land different, as "those who did not themselves subdue and cultivate the land had no right to prevent others from doing so."

Perhaps an ancient living tree standing in a ancient living forest would be a far brighter monument to the shared past of people and woodland, yet a disembodied slice of trunk displayed in a city park is a more truthful symbol of the culture that cut it. That a slice of ponderosa trunk is held so tenderly in its wood and granite cradle, honored with bronzed plaques, then marked with dates significant to human history—yet making no mention of the contributions and sacrifices made by forests to that history—speaks eloquently of the relationship to nature America has inherited from its European forbearers.

Perhaps a more equitable, if less revealing, display of our relationship to the natural world would be to instead mark at the center of that ponderosa the date on which English axes first entered the Caledonian Forest, then, a little further out, the date on which the last substantial stand of Scottish forest fell, then, following the ax westward and outward through the tree rings, the dates on which the first tree in New England and along the rivers of the Great Plains fell, and the first in the Rockies, the first along the Pacific shore, and then mark, somewhere out beyond the last ring and layer of bark—no more than a few inches into space perhaps—the date on which the last wild tree will likely fall to the ground and forever from human memory.

Still, there is hope. Mairi has learned, with nearly a decade of time with trees, to live among them, to even begin to enjoy shadowed places (I like to think that she is slowly recalling her own forest story). I, in turn, have learned to see a singular beauty in the emptiness of the Highlands, a beauty that comes from finding truths, if not trees. I've learned, from that storm-tossed land, of the consequences of forgetting. I've learned that the Highlands are not barren, but crowded with a forest of forgetting, a forest as important, perhaps, in understanding human nature, as the forest that preceded it. Hope lies in holding them both in memory.

From
Walden
Henry David Thoreau

Our village life would stagnate if it were not for the unexplored forests and meadows which surround it. We need the tonic of wildness—to wade sometimes in marshes where the bittern and the meadow-hen lurk, and hear the booming of the snipe; to smell the whispering sedge where only some wilder and more solitary fowl builds her nest, and the mink crawls with its belly close to the ground. At the same time that we are earnest to explore and learn all things, we require that all things be mysterious and unexplorable, that land and sea be infinitely wild, unsurveyed and unfathomed by us because unfathomable. We can never have enough of nature. We must be refreshed by the sight of inexhaustible vigor, vast and titanic features, the sea-coast with its wrecks, the wilderness with its living and its decaying trees, the thunder-cloud, and the rain which lasts three weeks and produces freshets. We need to witness our own limits transgressed, and some life pasturing freely where we never wander.

"A Letter for Montana Wilderness"
Rick Bass

Mike Espy
U.S. Representative
Dept. of Agriculture
U.S. House
Washington, DC 20515

Bruce Babbit
Secretary of the Interior
1849 C Street NW
Washington, DC 20515

Dear Mike Espy and Bruce Babbitt:

I lived in Mississippi from 1980 to 1987, where I worked as a petroleum geologist, and was active in the Central Chapter of the Sierra Club. I remember voting for you, Representative Espy, and being elated, and surprised, when you won. I have been impressed with your work in Washington, for Mississippi and all Americans, ever since. Secretary Babbitt, I was likewise thrilled with your appointment, and am hopeful you'll be allowed to help start reclaiming the public lands for the public.

I am living in northwestern Montana now, in a small valley called Yaak. It's 97 percent National Forest lands, under the jurisdiction, of course, of the U.S. Department of Agriculture. This week we had some friends visiting us from Mississippi who expressed dismay (a polite phrase) at the clear-cuts on the upper portion of this remote valley, the Yaak valley. The clear-cuts were up on the northernmost border, up where no one ever went. Hundreds of acres in size, the clear-cuts are on slopes so steep that the soil was washed away, leaving bare rock, where nothing can grow. Over twenty years after the logging, entire mountain faces are still totally bare, and will probably never revegetate. But the local timber industry got their cut out, and moved on.

It is a wild valley still, because it is on the U.S./Canadian line. It's lush and wet, rich in biodiversity, where it hasn't been clear-cut. It links the ecosystems of the northern Rockies to the Pacific Northeast—in that respect, it is the most strategic location in the West, with regard to the concept of conservation biology, and the flow and exchange of wild genes. It is home to grizzlies, wolves, wolverines, a woodland caribou, great gray owls, bull trout, sturgeon, and a host of other threatened and endangered species, and yet not *one acre* of designated federal wilderness exists in the Yaak.

Part of this is because it is not a "pretty" place for people to go and admire scenic vistas of ice-capped peaks; the forests that remain are either too thick or too fragmented (over ten thousand miles of roads, built into these last roadless areas at taxpayers' expense). These roads went first and fastest into the biggest, oldest forests, cutting the largest, most valuable trees for the international corporations. What's left isn't pretty, but it's wild; it's wilderness. It's where wild things live.

It rains a lot in Yaak. The very thing that keeps people away is what helps make this valley, despite its historical ravages from the local timber industry, a safe harbor for these wild, shy creatures. It's low elevation rain forest; it grows big trees.

The reason Yaak has not a single acre of designated wilderness is the power of corporate politics. Ten thousand people in Lincoln County used to vote one way: pro-mill. This is a small number of people and a small number of Montanans, but it's enough to have been perceived as a swing vote in Congressional politics, and some politicians in the West have a bad habit of controlling completely that public domain to which they are privy, and from which their constituents are so fortunate to receive the many benefits. It is my valley, my home, which the local mill, run by Champion International, has held hostage from the rest of the country for so many years.

But not any longer. I wish I could say it is because Congress has decided to protect the last of this country's federal wildlands. However, that's not the reason. The reason is that Champion, having cut almost all the forests, is leaving town—shifting operations down toward Mississippi, Arkansas, and other southern states, where the trees grow faster. They're coming your way, Representative Espy; be warned. They are not interested in preserving communities, or preserving anything.

The reason they're leaving is that in the late 1980s, to raise cash and reduce assets in order to defend against corporate takeovers that would pirate and fragment the company (in the same manner in which they had fragmented the land), Champion liquidated their assets as a defensive maneuver, to raise cash. Their assets, unfortunately, happened to be whole forests—not pine monocultures or "fiber factories," but complete riparian forests of larch, spruce, fir, and pine. In all instances, Champions' lands were adjacent to federal lands, and the wild creatures that moved through and among and above the forests did not understand, I'm sure, the murky betrayals of property boundaries. They only knew the land—their community—as a whole.

Over the course of only three years, Champion clear-cut almost all of their properties—867,000 acres in the state of Montana—most of it river bottom lands, the most critical wildlife habitat zone that exists. But before they wiped out all these forests, Champion made sure they were covered; they cut a deal with Congress, and with the U.S. Fish & Wildlife

Service, that would guarantee that Champion could continue cutting ad infinitum at this artificially high, hyperinflated volume; they obtained assurance from the Forest Service that after Champion ran out of all its logs, it could continue at the same pace on the federal lands. Never mind the damage to the ecosystem, or to the various species within that system; Champion wanted subsidizing, wanted exception from environmental regulations concerning endangered species, and wanted to keep, solely for profit's sake, the same steady (and too-high) volume running through the mill.

Because the Champion-owned mill in Libby would benefit in the short term by this increased, unsustainable volume of harvest, Champion presented themselves to the town of Libby, and Lincoln County, as purveyors of goodness, the cornerstone of civic responsibility: creators of jobs. (When actually they were robbing jobs by clear-cutting rather than the more labor-intensive selective cutting, and when they were sending raw logs from federal lands over to Asia, where the logs would be milled and manufactured and then sold back to American consumers as finished products at ten times the price at which they'd left the country. But because the taxpayers were paying, and because Champion was getting the free ride, it was all about volume and speed back then, not quality or sustainability. Champion knew all along they'd be leaving soon. . . .)

Environmentalists warned Libby that this was what Champion was up to, that Champion would screw the community the way they had screwed the land, but the political force of a small-town company town is well-known; the environmentalists were labeled (by Champion) as anti-growth and anti-American, and I regret to say that there is now yet one more instance of a boom-town gone bust. When I was a geologist in Mississippi I saw the price of oil plunge from $40 a barrel to below $10 a barrel; true boom-and-bust, but still even that is nothing compared to the way Champion went through the forests of Yaak before fleeing.

Watching it all has confirmed for me something I've always believed: that respect cannot be turned on and off, like electricity. Either you've learned to practice respect, or you haven't. I have been thinking that many of our problems in the American society stem from how we began, during the agricultural revolution, to use slaves—the voiceless minorities to control the land, and from there leapt hot-headed and wrong-hearted, already on the wrong course, or with false hearts, into the Industrial Revolution. Machines allowed us to treat the land with a newfound, careless lack of respect for forests and rivers and wild things, and this metastasized further into a lack of respect for all other voiceless minorities. Slavery was ended, but only on paper; we still had not learned respect. Or rather, had forgotten it.

In that sense—our culture, our heritage of dominating the voiceless minorities, the silent partners—I think that all of our problems are in one sense crises of the environment. I think that there is no separation, at heart's depth, between pollution, large-scale clear-cutting, homelessness, illiteracy, drug abuse, or poverty. They are all a crisis of disrespect. And I believe it is an epidemic: north, south, east, and west.

Right-wing critics have long sought to divide and conquer those who would argue for respect by attempting to pit environmentalists against other social activists, accusing the environmentalists of loving animals and forests more than people. (And in the meantime, the chainsaws rev louder, the bulldozers accelerate, and the mountains slide away into the muddy streams as the land developers, the land scrapers, call in their political favors and swing their three- or five-year projects onto the pub- lic lands, reap their profits, grow old, grow fat, and then die, fading from history and memory like the dying glow of a forgotten firefly, completely undistinguished in the annals of contributions to the human spirit and human dignity, instead leaving behind an anonymous legacy of ruined land and lost potential.)

We can't teach these kinds of people how to have respect for the voice- less; we can't muscle respect down their throats.

But if we can preserve some intact systems in the world where respect and logic still function in harmony—where actions still have conse- quences—and if people and communities can observe nature's cycles, both minute and grand, of respect, cooperation, and compromise—and if people can grow up learning to accept that there must be a few last wild places into which our furious domination will stop, a few last places into which we will not bull our way for profit—then I believe there is hope for other of our culture's ills of domination and loss of accountability. I believe that if we can learn to give up a tiny bit of power, only then can we learn to be strong. We were strong in our nation's beginnings because we were young, and because our country was blessed with an excess of natural resources—and now we must learn to be strong because we are old: to be wise, and not follow the bitter road to the very end of the last of those resources, now that they are so rare and precious—now that we have gone beyond sustainability.

I see these several years, on toward the end of the century, as the last chance we have to re-learn respect: for the land we live on, and respect for ourselves, and out of that, respect for all others. It may be romantic or naive but I am convinced that if a person can watch even a brief glimpse of unfragmented nature and all its woven connections—the way deer fawns are born the same week in early summer, for instance, that the grass is highest and greenest, to protect them from predators—then we can still hope to re-learn the forgotten notions of responsibility and accountability; of respect. In nature first and clearest we might be able to re-learn the oldest truth that everything is, and always has been connected; that a child's poverty in Mobile, Alabama, for example, robs from the security and strength of all of us.

The land got sick first, as a result of our dominating the voiceless minorities, and it is the land which must get well first, I think. I do not believe we can bypass our relationship to the land if we hope also to improve this culture's relationship to women, to children, to people of color, to the poor, the illiterate, the homeless. . . . We can't be good to each other if we can't even be good to a tree, or a forest.

It is not a question of priorities, as our foes would accuse us; it is not a question of jobs versus wilderness, or wild nature versus education for the hungry child in Mobile, or Portland, or Philadelphia. It is a question of finding strength once more—strength in wisdom and restraint, this time, rather than the fast-diminishing strength of aging muscle, or calamitous, misdirected passion.

LET ME RETURN TO MY VALLEY, please—the Yaak. Let me leave the abstract and return to the specific: this wild rainy place where there are not vistas, only clear-cuts and a few wild shy creatures, such as wolves and myself, living up here in a magical seam between the ecosystems of the flora-rich Pacific Northwest (*golden spleenwort, linear-leaved sundew, Queen's-cup bead lily*) and the magnificent predator-prey showcases of the northern Rockies: bear-elk, wolf-moose, cougar-deer, owl-grouse, eagle-trout, lynx-hare. . . .

Like no other place in the world, it is all still here, and all still connected, though just barely. We're down to five or six wolves, in this valley of almost 300,000 acres; down to nine or ten grizzlies. One caribou, occasionally, drifting back and forth among the lost no-man's lands between northern Idaho, Montana, and Canada. Twenty bull trout, and a handful of aging white sturgeon in the mighty Kootenai River.

I am not asking for tourists to flock to this wet, cold valley; unlike our other, prettier, rock-and-ice wildernesses, this one is not for humans.

What I'm asking is for action by Congress to help save the last hidden voiceless things up here.

Once again, the Yaak wilderness is not a wilderness for people to visit, or to attempt to thrash their way through the wet dripping ferns. They're not things to be seen, even if you could, which you can't—these few wintery mountaintop wolverines that drift back and forth across the Canadian line. What the greatest value this valley has to offer us is the way we choose to protect it—not the way we choose to tour or develop it. As many of us did for Alaska's wilderness in the 1980s, Yaak is a place to fight for until the fight is won, and then to know simply that here is a wild place we don't need to enter, that's not designed for our entrance—brushy jungle—a place where we demonstrated our strength by the manner in which we chose to give up our compulsive, chronic power over it, while the country still meant something: while there was still a bit of sweetness left to it.

I wrote this damn book about this valley, some time ago, back when I was naive; I referred to the valley by its proper name, Yaak, rather than some made-up name of fantasy. For whatever reasons, the book struck chords in the hearts of readers, of which there have been many. It was a story about starting a new life, and noticing new things, in a new place.

I've been up here seven years. It used to be close to Eden; up until I wrote that book, there was rarely a car on the road. People rode horses down the road. You could walk down the road and see a mountain lion napping in the dust in a patch of sun. Grouse, too, would come out and dust-bathe along the road, or peck gravel for their gizzards; sometimes a hawk would plunge from the sky and nail one, right before your eyes. If a snowshoe hare ran out in front of you, you paused half-a-second; chances were good that a coyote, or bobcat, or even a wolf would be hot on its heels. Badgers sunned themselves dry on the roads after late-day summer thunderstorms had filled their holes briefly with water. Bears padded across the roads, quickly, furtively.

Since my book, all that's changed. The animals are still out there, but they're wary, because there are so many more cars now. A good number of the drivers of those cars come searching for me, uninvited, to knock on my door and interrupt the living dream of my life, the waking dreamtime, to tell me they read the book and liked it (or didn't). That's my problem, and my due, I suppose. And perhaps it's best that the animals have learned to move away from the roads; perhaps this increased wariness will serve them better, in the future. Still, I feel a sense of loss, of unnecessary lost innocence.

This is not to say that there are not still infrequent surprises from out of the blue, when animals reveal themselves. A great blue heron spearing tiny brook trout; a golden eagle among thirty ravens, sitting atop a road-killed deer. One winter morning my wife and baby daughter and I looked out the window at the pond to watch a family of five otters playing on the pond ice at twenty below, diving in and out of a hole in the ice, catching fish and playing tag. Yesterday evening, a friend drove up to visit, and as we sat there talking until dusk, a badger crawled under his car, perhaps to lick antifreeze from beneath his slightly-leaking radiator—his car had overheated while chasing elk poachers earlier in the afternoon—and the badger charged us and my dogs (the same dogs that had been torn up by a pack of coyotes several weeks earlier). The badger then retreated to beneath the car and wouldn't leave, and wouldn't let Jesse or myself near the car. It growled and hissed and snarled whenever we got within twenty yards of it, and we had to sit on the porch and wait until after dark before the badger waddled away, flowing silver across the yard under a perfect full moon. . . .

What is the point? The point is that here is a minority—the minority of the wild, the minority of the untamed, the uncompromising—and that it is in our power to give back its freedom—to allow the last two roadless areas—Roderick/Grizzly, and Mt. Henry/Pink Mountain/Gold Hill—to remain what they are, and to become what they will, under their own design, by their own processes. Generosity, Barry Lopez has noted, is an act or courage. Protecting one of the last two areas is not enough; we need to protect both of them for there to be a weave, a relationship of the wild.

The valley to the east of here, the Ural Valley, once home to grizzlies and wolves, has disappeared completely—was flooded by the building of Libby Dam, which generates electricity and sends it to Tacoma, which then sells it via the Bonneville Power Administration directly to California.

Seventy miles to the south, the world's largest copper and silver mine is proposed to tunnel beneath the Cabinet Mountains Wilderness. The mining company (Canadian-owned Noranda), the U.S. Forest Service, and the U.S. Fish & Wildlife Service have broken up time-honored family home ranges of grizzly bears, the historic and ancient cultures of grizzly families, into arbitrarily-numbered compartments called Grizzly Bear Habitat Management Units—GBHMUs—and like little children bullying something smaller or more voiceless than themselves, the agencies claim to always have habitat available to the bears, *somewhere*, at a 70 percent habitat effectiveness rating. (When I went to school, 70 percent was a D-minus.)

Furthermore, the agencies shuttle these effectiveness ratings around every three years, via logging sales, and expect that these bear families, these bear cultures, will find their way from Compartment 1 to Compartment 37, and then when Compartment 37 is logged, on to Compartment 62. . . . The Forest Service calls this "displacement habitat," and feels clever, that they'd "provided" this displacement area (usually at the 70 percent level). It reminds me eerily of our government's shuttling of Native Americans from reservation to reservation.

Grizzly cubs stay with their mother for at least three years, so in effect, there will be grizzlies that are asked to move every time they become pregnant, and even worse, subadult bears will be expected to find their way from piss-poor Compartment 1 to piss-poor Compartment 37 all on their own, in their first wobbly year of independence. This is pseudo-biology, computer manipulation used to continue stealing from the public lands; it hides from the U.S. Congress and the taxpaying public the biological tenet that grizzlies live, or try to live, their entire life in a single home range. They pass on information about that one place—their *home*—to their offspring: where to find berries, when to hibernate, where to hibernate, where to hide. . . . There's no word for it other than *culture,* and the Forest Service is pretending this doesn't happen. This is a deception of Congress and the American public; the culture of bears is being destroyed. Activities on the National Forests are destroying the bears themselves, just as killing all the bison a hundred-plus years ago finished off the culture of Native Americans. If the last two roadless cores in Yaak aren't protected, the grizzlies (and a whole lot of other species) aren't going to survive. It's simply not going to happen. We might as well just shoot them in the head and get it over with.

In this wild, unpeopled valley far from your homes, up on the Canadian/Idaho/Montana border, lies the clean but disappearing chance to keep a balance, and to preserve a working model, an example of a system that's still based on logic and the cooperative integration of every single element in that system, large or small, voiceless or bold. This, I believe, is nature's definition for the word *compassion,* which I'm afraid is a thing we move further and further from yearly.

I do not mean to sound like a salesman, as if I'm recommending preservation of wilderness for the sake of slowing our descent down the wrong turn that our culture (like so many ancient, extinct cultures before us) has taken. I believe it's much simpler than that: that we should preserve our last tiny remnants of this northernmost wilderness in the Lower 48—not just Roderick Mountain, but Grizzly Peak and Mt. Henry, Pink Mountain and Gold Hill, in this cornermost, forgotten valley—because it is simply the right thing to do.

These days we no longer have the comfort of hiding out in a moral middle-ground, a no-man's land of no decisions. These days—down to our last one percent of unroaded American wilderness—if you are not acting to save it, then you are aiding in its development. Every one of us must line up either on the side of compassion, or on the side of manipulation, control, and domination.

I refuse to believe these latter ugly things are hard-wired into our existence. I continue to hope and believe we have within us the ability, the grace, to finally learn a light touch—to learn to lift our hands from a thing and step back and allow it to take its own course, to choose its own way, at its own pace. If we can't do that on such a small area as these 55,000 acres, the last we have, what chance do we have of a similar grace returning to and sweeping across the whole country?

Let anyone who reads this plea please act to save a long-ignored, long-abused voiceless place: to write letters for these last Yaak mountains without ceasing. Protecting Roderick alone as wilderness is not enough, anymore than being compassionate half the time is enough; this would only create an island, rather than protect a woven, breathing system of grace. We need to protect this last *system*.

As ever, there will be some who will oppose the notion of protecting the last wilderness we have—a thing that won't ever be coming back. There will be a few who will want to build roads into and through and over these last four mountains in Yaak, after having warmed up on the first ninety-six. Even if you don't recognize these people at first, you will later. They're the ones who aren't planning to stick around. They're the ones who are in a hurry, on their quick way to exercise their unlearned dominance over the next voiceless population, and they certainly will not stop at trees and animals. You will recognize them after they have gone on into history, and have taken with them a thing that might otherwise have remained beautiful, whole, and that rarest of things, free.

Thank you for your time—for listening.

Sincerely,
Rick Bass

cc: George Miller, Chairman, House Committee on Natural Resources, National Parks, Forests, and Public Lands Subcommittee, and Bruce Vento, Subcommittee Chairman
> U.S. House
> Washington, DC 20515

cc: J. Bennett Johnston, Chairman, Senate Energy and Natural Resources Committee, Public Lands, National Parks, and Forests Subcommittee, and Dale Bumpers, Subcommittee Chairman
> U.S. Senate
> Washington, DC 20510

Jack Ward Thomas	Regional Forester
U.S. Forest Service	Region One,
Box 96090	U.S. Forest Service
Washington, DC 20090	Box 7669
	Missoula, MT 59807

Max Baucus,	Conrad Burns,	Pat Williams,
D-Montana	R-Montana	D-Montana
U.S. Senate	U.S. Senate	U.S. House
Washington, DC	Washington, DC	Washington, DC
20510	20510	20515

Also in the House (Washington, DC 20515): Subcommittee members Ed Markey (MA), Nick Rahall (WV), Peter DeFazio (OR), Tim Johnson (SD), Neil Abercrombie (HI), Carlos Romero-Barcel (PR), Karan English (AZ), Karen Shepherd (UT), Maurice Hinchey (NY), Robert Underwood (GU), Austin Murphy (PA), Bill Richardson (NM), Patsy Mink (HI), Jim Hansen (UT), Bob Smith (OR), Craig Thomas (WY), John Duncan (TN), Joel Hefley (CO), John Doolittle (CA), Richard Baker (LA), Ken Calvert (CA), Jay Dickey (AR), Gerry Studds (MA), Kika de la Garza (TX), Carolyn Maloney (NY).

Also in the Senate (Washington, DC 20510): Subcommittee members Bill Bradley (NJ), Jeff Bingaman (NM), Daniel Akaka (HI), Richard Shelby (AL), Paul Wellstone (MN), Ben Nighthorse-Campbell (CO), Mark Hatfield (OR), Robert Bennett (UT), Arlen Specter (PA), Trent Lott (MS).

President Bill Clinton
The White House
1600 Pennsylvania Ave.
Washington, DC 20500

Vice President Al Gore
Office of the Vice-President
Old Executive Office Building
Washington, DC 20515I

❧

"Future Madness"
John A. Livingston

No doubt most if not all contributors to this volume will have commented in some form at some time upon the global "development" idea. Having myself delivered a quite sufficient number of harangues on that subject, here I shall dwell instead upon two symbiotically intertwined concepts which serve to support the "development" imperative. The twin symbionts are (a) sustainability and (b) the future. Both are cultural and ideological artifacts, abstract social fabrications.

Let us begin with sustainability. What is it that is to be sustained? What is it that we are to maintain, prolong, nourish, and support indefinitely? If, as it appears to be, it is the advancement of the human monoculture, that is straightforward enough. Monocultures do two things—they homogenize and they simplify. Everyone knows that homogenization and simplification are destructive of natural communities. The human monoculture is death to Nature. That is the implicit end of the "development" of the biosphere. We are being instructed to sustain the explicit means.

As a naturalist I have always believed that the biosphere was already fully developed long before we began to homogenize and simplify. At any point in its career, the biosphere was and is fully developed. It is now, it was when some space junk blew away the dinosaurs, it was when the first crossopterygians came ashore, and it was in the beginning. I have never understood the perceived necessity of the biosphere being "improved" in the image of a single species, especially one so grotesquely maladjusted, in an ecological sense at least, as our own. That is how I comprehend the advancement of the human monoculture. That is what we are being entreated to sustain.

But maybe that's not it. Maybe what is to be sustained is not business as usual, but rather some revised, scaled-down version of the human project. Maybe we should facelift the biosphere not in the image of the urban–industrial monster but rather in the image of the shepherds and the shepherdesses gamboling on the greensward in some New Age Arcadia.

That would be nice. There would be sheep, and there would be music, and there would be sunshine and puffy cumulus clouds—just little ones, mind you—and all would be in harmony with Nature. Trouble is, any really heavy-duty gamboling eats up a hell of a lot of greensward, and what with the sheep and all, there wouldn't be room on the planet for six thousand million frolickers. Let alone any gardens or vineyards. Or anything else, for that matter. So I reckon it's not that either.

Maybe what is to be sustained is renewable resources. The term "renewable resource" has itself gained full-blown oxymoronic status. Codfish, great whales, old growth forests, tropical soils, fresh water aquifers—even frogs and songbirds—all used to be called renewable resources. We need not invest much cerebration on that one. Unless there's some way to cash in on Great Lakes zebra mussels, which renew themselves with quite astonishing fruitfulness—and without the help of any resource management expertise whatever.

I suppose the ultimate beneficiary of sustainability—at least rhetorically—would have to be life process itself. We need not worry about that one either. Life process will go on—with us, or without us. That, by the way, is to be construed as a positive statement.

But still we have the sustainability imperative. In whose interest are we to sustain whatever it is that we are to sustain? Nature's interest? Not bloody likely. It must be the interest of the industrialized nations. Or is it that of the would-be industrialized nations? Or of simple human fecundity?

Or, maybe it's the interest of an even higher purpose. I think we're getting warmer. Maybe what is to be sustained is not a concrete tangible but an idea. An abstraction of some kind. Maybe what we are being asked to maintain is the very concept of the future.

We thrash about in a maelstrom of ironies. One of the more crashing is our collective pride (hubris, more like) in our abstraction. Other animals regularly abstract in their daily lives (problem-solving may be universal in vertebrates at least, and there is plenty of evidence from octopi, among others), but the human does very little else. Abstract reasoning is the human specialty. But specialization, like the free lunch, always has its cost.

Along our improbable evolutionary way, abstraction seems to have become increasingly dominant relative to the rest of our "normal" animal gifts. So important did it become that at some point a qualitative Rubicon was crossed, and we became no longer the masters and possessors of abstraction, but its creatures.

We moved into a state of total dependence and servitude, most especially to instrumental (technical) reasoning, and in so doing lost our participatory relatedness to and identification with the wider life community. No longer able to perceive the rest of the multitudinous biosphere in any way other than the abstract and utilitarian, we had become ecological loose cannons, dangerous misfits in what had been theretofore an ineffably good, perfectly functioning and integrated community of being. We found ourselves alone.

We rationalized our loneliness as uniqueness. That was pretty wide off the mark. All species, all individuals are unique. So we rationalized our loneliness as ordained separateness, prefigured noblesse. God said that there should be apartheid on Earth. We liked that, so we wrote it down, and codified it. So went the reasoning—or, as I prefer, the rationalization.

There is a monumental irony in our specialization, and that is in the abstraction of time. More precisely, the abstraction of the future. Obviously, all technical or instrumental thinking and tinkering, and all "progress" orientation requires the prior assumption not only of perfectibility but also of goals, and the assumption of goals in turn necessitates the prior abstraction of the future. Forward-thinking by most well-adjusted animals appears to be confined to the duration of the task at hand—the strategy of the hunt, the best approach to the immediate challenge, the weighing of present options in the here and now.

Non-human beings and a shrinking number of human societies have always lived in the joyous fullness of the present. Our society has utter contempt for their apparent inability or unwillingness to be victimized by morbid obsession with what may or may not obtain tomorrow.

Our future-orientation is virtually total, certainly at least in Western cultures. Our self-identities as individuals and societies and civilizations—maybe even as a species—appear to be cast in terms not of what we have been, not of what we are, but of what we shall become. But remember that only humans become. Non-human beings may be seen as evidence of the creativity of natural selection, but humankind must be seen differently. Our species must be seen as the eventual and final culmination and realization of life's purpose. The future of planet Earth is the human future.

A difficulty arises in the fact that by definition the future never comes. It follows that in practice we can never become. Our self-identity can never be found. Having long since rejected our self-identity as Nature, we chose to put all our chips on the future, whose margin, like that of Tennyson's struggling Ulysses, fades forever and forever as we move. No matter; the priests and the secular gurus, the technocrats and the humanists generally, keep flogging our progress toward future self-realization.

But if we identify solely with the future, then surely we never grow up. This is, of course, the common condition of all domesticated mammals, among whose number I have recently argued the human species must be included. Paul Shepard has shown that Western society itself suffers from arrested development. (That was in 1982; in the 1990s perhaps we must say that Western society is developmentally challenged. Which indeed it is, in more ways than one.) As individuals, societies, and civilizations, we are forever becoming, forever questing for mature adulthood.

It follows that, in the advancement of our quest, whatever must be done, must be done. If it means the simplification and homogenization of natural communities and processes, so be it. If it means the extirpation of whole floras and faunas—without replacement—then that is in the nature of things. The human enterprise is necessary.

The human future is imperative. The subsidy Earth and earthly non-human beings are required to pay for human immortality is of course factored out, not in. That is an externality—an incalculable, as the technocrats like to say. Regrettable, perhaps, but necessary and inescapable.

My final question concerns how the future, however imagined, is to be sustained. The idea of the future we sustain by mantra, litany, catechism, ideology, casuistry, and a good shot of group self-mesmerism, to say nothing of human chauvinist narcissism. The realization of the future—the necessary advancement of the human monoculture—is to be sustained by Nature.

The clear assumption is that Nature owes us. It is Nature's appointed task—its reason for being—to maintain and nourish the human project. Nature was provided to serve the Chosen Species. That is the received cultural and historical wisdom that sustains the ideology of the necessary primacy, among all things, of the human enterprise.

I do not know whether all forms of madness need inevitably be tragic. But this variety has all the earmarks.

🦌

"The Good News"
Gary Lawless

Roads disappear, and the caribou wander through.

The beaver gets tired of it, reaches through the ice, grabs the trapper's feet,

pulls him down.

Wolves come back on their own, circle the statehouse, howl at the sportswriters,

piss on ATVs.

Trees grow everywhere.

The machines stop, and the air is full of bird song.

Part Three

Wildness, Philosophy, and Spirituality

"Nature is the art of God."

—Dante

The concern for wilderness—and for its essential trait, wildness—now welling up in society transcends both biological and political considerations. "An identification with the wild," wrote George Sessions, "has increasingly come to characterize the ecocentric perspective and a fully developed ecological consciousness." For many, the issues of wildness and wild landscapes are, at core, spiritual and ethical, and the biological savvy exhibited by such a large segment of the activist community tends to come after their having been drawn into activism by love of Nature, personal ethics, native intelligence, and gut instinct.

From
"The Stranger's Ways"
J. Donald Hughes

Two completely different ways of seeing the earth and treating the earth have existed side by side in America for the past few centuries. American Indians and European-Americans lived in the same land, but neither clearly understood the attitudes of the others toward the land, or what the others thought they were doing with it and with the forms of life and natural resources on it. As the great Nez Piercé, Chief Joseph, said, "We were contented to let things remain as the Great Spirit made them. They were not and would change the rivers and mountains if they did not suit them." The two peoples really lived in different worlds, since human beings experience the world in terms of their own perceptions and points of view. Though each spoke to the other, neither realized how different the ideas were that underlay the words used by the other.

The people who came across the Atlantic carried with them a traditional European view of the natural environment, which denied that there is any spirit in nature, or anything intrinsically valuable about the things in nature—and they did refer to natural objects and life forms as "things," never "persons" or "beings." Therefore they felt that they could use any creature in the natural environment without concern for the creature itself or its role in natural systems. Only its utility to human beings, especially, or perhaps only, to the individual owner of the land where it was found, was considered. This attitude is still dominant today. Non-Indians often assume that Indian conservation practices must be a result of economic necessity or expediency, because non-Indians characteristically think in those terms themselves. They do not share the traditional Indian view of the world as a place where spirit power, ritual, and reciprocity with animals and plants are operative.

❧

From
Ecological Revolutions:
Nature, Gender, and Science
in New England
Carolyn Merchant

Mechanism and the Domination of Nature

The mechanistic philosophy developed by the natural philosophers of seventeenth-century Europe legitimated the capitalist revolution and its domination of nature. Mingled with the rhetoric of the Great Chain of Being, Mother Earth, and the Garden of Eden that focused New England thought in the eighteenth century had been an undercurrent of instrumental concepts that would structure the management of nature in the nineteenth century. Mechanical metaphors and the rhetoric of manifest destiny became core concepts of a modern philosophy that saw the world as a vast machine that could be mathematically described, predicted, and controlled. A new chemical paradigm would quantify associations and dissociations of elements in soils and plants so that yields and profits could be predicted and increased. Mother Nature was delivered to the laboratory to undergo scientific experimentation.

In the seventeenth century, Descartes had characterized the human body as a machine and compared it to a clock that operated according to mechanical laws, while English political theorist Thomas Hobbes (1588–1679) had described the body politic in terms of springs, strings, and wheels. Newton's *Mathematical Principles of Natural Philosophy* had integrated celestial and terrestrial mechanics into a cosmology founded on the certainty of mathematical law and the divisibility of matter into inert atoms moved by external forces. Mechanism thus offered a unified theory of the human body, society, and the cosmos.

The mechanistic construction of nature is based on a set of ontological, epistemological, methodological, and ethical assumptions about "reality." First, nature is made up of discrete particles (atoms or later subatomic particles). Second, sense data (information bits) are discrete. Third, the universe is a natural order, maintaining identity through change, and can be described and predicted by mathematics. Fourth, problems can be broken down into parts, solved, and reassembled without changing their character. And fifth, science is context-free, value-free knowledge of the external world. As constructed by the seventeenth-century "fathers" of modern science, the mechanistic model served to legitimate the human prediction, control, and manipulation of nature.

During the eighteenth century, Enlightenment elites appropriated from the scientific idiom metaphors that described human society in terms of instruments, machines, gears, pulleys, and balances. The homes of gentlemen exhibited pendulum clocks, cannonball escapements, orreries (working models of the solar system), music boxes with moving figurines, telescopes, microscopes, meteorologic instruments, pulley systems for raising food or opening doors, mechanically driven fountains, hydraulic devices whose sudden commencement surprised and startled garden visitors, and a host of other machines and playthings. Thomas Jefferson's home at Monticello was filled with copies of European models as well as with inventions of his own. In the 1800s New England clockmakers produced large numbers of shelf clocks, wall clocks, and grandfather clocks with second hands and rocking manikins or ships that indicated whether the clock was operating.

Mechanical language derived from European philosophers permeated the writings of the nation's founding fathers and their fundamental documents—the Declaration of Independence and the Constitution. Scottish philosopher Adam Smith (1759) had explained how "all the several wheels of the machine of government [might be] made to move with more harmony and smoothness, without grating upon one another, or mutually retarding one another's motions." Referring to his role in the adoption of the Declaration of Independence, John Adams hoped that he had been "instrumental in touching some springs and turning some small wheels" of historical development. The language of the Declaration itself was Newtonian in emphasizing the necessity for action based on observation of a sequence of events. James Madison's "Federalist 10" (1788) treated society as a balance among atomized factions. The Constitution of the United States (1789) was constructed as a system of balances among the powers allocated to the three separate elements of government.

Mechanical metaphors also began to infuse the popular literature of eighteenth-century America. Since their inception, almanacs had taught

Copernican astronomy and Newtonian science to their readers. But a mechanical philosophy going beyond instruction on scientific laws extended the description of nature as machine. Nathaniel Ames informed the readers of his *Almanac* in 1754 that the Divine Artificer initially had made the body of man "a machine capable of endless duration"; but after Eve's ingestion of the forbidden apple, the living principle within had fallen into disharmony with the body, disrupting the smooth functioning of the parts.

Similarly, Ames's son and successor, Nathaniel, described the human body a decade later as "an infinitely more curious machine or piece of clockwork than anything contrived by man." Just as a clockmaker should know the make and machinery of the clock before attempting to repair it, a doctor should know the "make and machinery of the body." The physician must set his reason to work to "find out which of the pipes, springs, or strainers is out of order . . . whether they want stiffening or loosening, oiling or cleaning."

The mechanical paradigm of association and dissociation of atoms provided the rationale for agricultural and social improvement, while males moving in free association became the entrepreneurial pattern. The westward movement encouraged spatial mobility; emerging capitalism promoted social mobility. Manufacturers produced machines made of interchangeable parts, while managers hired wage laborers as replaceable cogs in the machinery of production. Men organized businesses for the specific purpose of profit making, and each male was free to associate with or to dissociate from them as opportunities arose. Efficiency dictated that each remain a part of a company or operation only as long as it was profitable and then be ready to move onward or upward to a new venture. Productivity and profit were the deciding factors rather than emotive bonds between individuals.

The rhetoric of manifest destiny sanctioned the spatial motion that encouraged control over natural resources as Europeans swept westward bearing the torch of "civilization." The elder Ames foresaw the day when art and science would change the "face of nature" west of the Appalachians as far as the Pacific Ocean. As civilization moves across the western deserts, he wrote in 1754, "the residence of wild beasts will be broken up, and their obscene howl cease for ever." Instead, rocks and trees would dance to the music of Orpheus, and gold and silver treasures would be discovered in barren rocks long hidden from "ignorant aboriginal natives." Iron ore already dug in the East would be set to practical use in creating plowshares and swords. All this would be sanctioned by God and the Gospel as the heathens were dispelled by light.

In extolling the present superiority of Christianized European civilization, Ames sent forth the ripples of manifest destiny that would legitimate the nineteenth century's westward movement: "the progress of human literature (like the sun) is from the east to the west; thus has it traveled thro' Asia and Europe, and now is arrived at the eastern shore of America." He concluded with a prophetic message to those of us who live today: "Ye unborn inhabitants of America! . . . when your eyes behold the sun after he has rolled the seasons round for two or three centuries more, you will know that in anno domini 1758 we dream'd of your times."

Ames had anticipated by almost one hundred years Missouri's Thomas Hart Benton, whose famous 1846 address to the Twenty-ninth Congress justified American expansion to the Pacific. Using the Bible as legitimation for manifest destiny, he proclaimed that the white race had "alone received the divine command to subdue and replenish the earth: for it is the only race that . . . hunts out new and distant lands, and even a New World to subdue and replenish." By then (as Ames had predicted), the red race had almost disappeared from the Atlantic. "The van of the Caucasian race," Benton gloried, "now top the Rocky Mountains, and spread down on the shores of the Pacific." Inevitably, white influence on the 400 million people comprising the yellow race of Asia would be felt, "a race once the foremost of the human family in the arts of civilization" but now grown degenerate. Under the influence of white trade and marriage, the sun of civilization would once again shine on them.

And John Quincy Adams, that same year, in promoting American expansion into the rich farmlands of Oregon, quoted from Genesis 1, urging the young nation to "make the wilderness blossom as the rose, to establish laws, to increase, multiply, and subdue the earth, which we are commanded to do by the first behest of God Almighty."

Upward mobility was to be achieved through extraction of natural resources from the earth by the most efficient and profitable method. In 1844, Yale's Whig essayist Calvin Colton characterized America as a "country of self-made men" and the American environment as a source of unlimited natural resources. "Providence has [given] us a rich, productive, and glorious heritage. . . . The wealth of the country is inexhaustible, and the enterprise of the people is unsubdued. . . . Give them a good government, and they cannot help going ahead, and outstripping every nation on the globe."

Echoing the rhetoric of upward mobility, agricultural improvers exhorted farmers to become rich, like the self-made men of other classes, by calling upon the resources of the earth mother. "When he would add to his earthly treasure," advised Reverend H. M. Eaton in his speech to the farmers of Kennebec County, Maine, "he draws from the boundless resources of wealth concealed in the bosom of *mother earth*. She has a treasure for the farmer of such a nature, and so vast in extent, that giving does not impoverish her, and by withholding she is not enriched."

Subduing the Earth

To the biblical mandate for dominion over nature that had guided the Puritan transformation of the environment, nineteenth-century science added mechanical and chemical methods for altering it. A harmonious fusion between the Bible and Baconian instrumentalism established science as the method to be used in subduing the earth to improve the human condition. Francis Bacon's *New Atlantis* (1627) had been dedicated to the alteration of nature through scientific instruments and experiments. Fully compatible with the mechanistic view of nature, his scientific research program sought to command nature by obeying its laws. Grounded in the mechanical and chemical paradigm and augmented with an array of new mechanical technologies, Baconian utilitarianism became the ethic of agricultural improvement.

Improver Henry Colman began his address to the Hampshire, Franklin, and Hampden Agricultural Society of Massachusetts by quoting Bacon's principle: "The effort to extend the dominion of man over nature, is the most healthy and most noble of all ambitions." Just as "the great master of philosophy" had linked control over a female earth to the recovery of the garden that Eve had lost for humankind, Colman characterized the earth as a female whose productivity would help advance the progress of the race. Agricultural improvement, he believed, was the most salient example of how human power and creativity could help in controlling nature. "Here man exercises dominion over nature; . . . commands the earth on which he treads to waken her mysterious energies . . . compels the inanimate earth to teem with life; and to impart sustenance and power, health and happiness to the countless multitudes who hang on her breast and are dependent on her bounty."

Speeches made to farmers at local societies for promoting agriculture routinely drew on female earth rhetoric. M. B. Bartlett in his address to the farmers of West Oxford County, Maine, eulogized the pioneer struggles "with mother earth" that had created "sinewy arms and brawny chests" in the transformation of the cold, granite bedrock of New England into a habitable climate. Until only recently the wolf had prowled on the outskirts of town and bear were encountered in the forests; older farmers recalled Indian attacks and scalpings. Those who had gone west, Bartlett admonished, should have stayed to complete New England's transformation into a garden. He exhorted those who remained behind to finish the struggle: "Force the earth to yield to you her hidden wealth, act out your destiny with all the force and goodness that is in you."

As God's agent, echoed Ezekial Holmes, the farmer had submitted nature to "repeated and successive blows of the axe, hewing out, as it were, a farm and a homestead from Nature herself . . . making the wilderness blossom as the rose, and . . . converting the lair of the wild beast into smiling farms and thriving villages."

The transformation of the "howling wilderness" into "fruitful fields" and "smiling farms" was pushed with such frequency in journal articles, speeches, and commentaries as to become commonplace. John Goldsbury, writing for the *New England Farmer* in 1855, queried whether the descendants of Adam and Eve could have obeyed God's command to "be fruitful and multiply and replenish the earth and subdue it" without cultivating the land, even if they had retained their abode in the Garden of Eden. Human labor coupled with the progress made in agriculture, mechanics, science, and art had changed New England to prosperous towns of happy, enterprising people. "Agriculture is the mother of some and the nurse of all the mechanic arts," John Bullard told the Western Society of Middlesex Husbandmen in 1803. By cutting down the trees and clearing the land of rubbish, the earth could be made "an agreeable . . . abode to the children of men." Once the fields had been tilled and planted the farmer could delight in the valleys of corn, the pastures filled with sheep, and the perfume of the fields.

Like Bacon, Sidney Perham reminded Maine's Oxford County Agricultural Society that "knowledge is power." He advocated that the farmer follow and learn from nature's laws rather than subjugating it; but in true Baconian spirit, he also advised that this be done through scientific experiment. "We must enter her laboratory, and learn her various modes of distributing and combining her elements for the production of given results; and then we shall find her a co-worker with us."

While the ethic of the animate cosmos had urged farmers to imitate nature and hasten its own processes, the mechanical paradigm offered new techniques and powerful machines to fundamentally transform it. Dr. N. T. True, addressing Maine's Cumberland County Agricultural Society, urged farmers to appropriate nature's own processes wherein "instead of suffering the land to go fallow . . . she makes use of a rotation in crops." In planting pines on worn-out lands, nature provided deep roots that brought up potash and other nutrients. Deciduous trees used the potash in their leaves, returning it to the soil when they fell. "You will find nature slowly, but surely at work, forming a suitable soil for some other crop, which in the lapse of ages she may see fit to introduce."

But in rotating crops and improving the land, True advised the farmer to use machines. He should hoe and shell corn, mow and rake hay, reap, thresh, and winnow grain, pull stumps, saw wood, pare apples, and churn butter with the new agricultural machinery. He should plow deeper, manure more heavily, and cultivate better than had his father and grandfather. With crop rotation, he need no longer blame his "good mother, earth" for failed harvests. "Be kind to your mother," pleaded True, "and she will always be kind to you in return. . . . The more carefully we study her in her works, the more probable will be our success in our attempts at imitation."

Nature was the best teacher when it came to managing her woodlots for shipbuilding timber. Even if the back forty had been improvidently hacked for firewood or exterminated in conflagrations, "Kind Nature, man's best friend, attempts to repair these breaches in her sylvan shades." She planted seedlings of oak, pine, ash, maple, and many other useful trees important for a "commercial people." But if thwarted by turning out cattle and sheep to graze, she would retaliate with thistles, burdocks, and brambles. Yet even as elites offered their allegiance to the earth as mother and teacher, their mechanical, instrumental view of nature was subtly legitimating and advocating its management and manipulation.

From
Man and Nature
George Perkins Marsh

Man has too long forgotten that the earth was given to him for usufruct alone, not for consumption, still less for profligate waste. Nature has provided against the absolute destruction of any of her elementary matter, the raw material of her works; the thunderbolt and the tornado, the most convulsive throes of even the volcano and the earthquake, being only phenomena of decomposition and recomposition. But she has left it within the power of man irreparably to derange the combinations of inorganic matter and of organic life, which through the night of aeons she had been proportioning and balancing, to prepare the earth for his habitation, when, in the fullness of time, his Creator should call him forth to enter into its possession.

Apart from the hostile influence of man, the organic and the inorganic world are, as I have remarked, bound together by such mutual relations and adaptations as secure, if not the absolute permanence and equilibrium of both, a long continuance of the established conditions of each at any given time and place, or at least, a very slow and gradual succession of changes in those conditions. But man is everywhere a disturbing agent. Wherever he plants his foot, the harmonies of nature are turned to discords. The proportions and accommodations which insured the stability of existing arrangements are overthrown. Indigenous vegetable and animal species are extirpated and supplanted by others of foreign origin, spontaneous production is forbidden or restricted, and the face of the earth is either laid bare or covered with a new and reluctant growth of vegetable forms, and with alien tribes of animal life. These intentional changes and substitutions constitute, indeed, great revolutions; but vast as is their magnitude and importance, they are, as we shall see, insignificant in comparison with the contingent and unsought results which have flowed from them.

The fact that, of all organic beings, man alone is to be regarded as essentially a destructive power, and that he wields energies to resist which, nature—that nature whom all material life and all inorganic substance obey—is wholly impotent, tends to prove that, though living in physical nature, he is not of her, that he is of more exalted parentage, and belongs to a higher order of existences than those born of her womb and submissive to her dictates.

There are, indeed, brute destroyers, beasts and birds and insects of prey—all animal life feeds upon, and, of course, destroys other life—but this destruction is balanced by compensations. It is, in fact, the very means by which the existence of one tribe of animals or of vegetables is secured against being smothered by the encroachments of another; and the reproductive powers of species, which serve as the food of others, are always proportioned to the demand they are destined to supply. Man pursues his victims with reckless destructiveness; and, while the sacrifice of life by the lower animals is limited by the cravings of appetite, he unsparingly persecutes, even to extirpation, thousands of organic forms which he cannot consume.

The earth is not, in its natural condition, completely adapted to the use of man, but only to the sustenance of wild animals and wild vegetation. These live, multiply their kind in just proportion, and attain their perfect measure of strength and beauty, without producing or requiring any change in the natural arrangements of surface, or in each other's spontaneous tendencies, except such mutual repression of excessive increase as may prevent the extirpation of one species by the encroachments of another. In short, without man, lower animal and spontaneous vegetable life would have been constant in type, distribution, and proportion, undisturbed for indefinite periods, and been subject to revolution only from possible, unknown cosmical causes, or from geological action.

But man, the domestic animals that serve him, the field and garden plants the products of which supply him with food and clothing, cannot subsist and rise to the full development of their higher properties, unless brute and unconscious nature be effectually combated, and, in a great degree, vanquished by human art. Hence, a certain measure of transformation of terrestrial surface, of suppression of natural, and stimulation of artificially modified productivity becomes necessary. This measure man has unfortunately exceeded. He has felled the forests whose network of fibrous roots bound the mould to the rocky skeleton of the earth; but had he allowed here and there a belt of woodland to reproduce itself by spontaneous propagation, most of the mischiefs which his reckless destruction of the natural protection of the soil has occasioned would have been averted. He has broken up the mountain reservoirs, the percolation of whose waters through unseen channels supplied the fountains that refreshed his cattle and fertilized his fields; but he has neglected to maintain the cisterns and the canals of irrigation which a wise antiquity had constructed to neutralize the consequences of its own imprudence. While he has torn the thin glebe which confined the light earth of extensive plains, and has destroyed the fringe of semi-aquatic plants which skirted the coast and checked the drifting of the sea sand, he has failed to prevent the spreading of the dunes by clothing them with artificially propagated vegetation. He has ruthlessly warred on all the tribes of animated nature whose spoil he could convert to his own uses, and he has not protected the birds which prey on the insects most destructive to his own harvests.

Purely untutored humanity, it is true, interferes comparatively little with the arrangements of nature, and the destructive agency of man becomes more and more energetic and unsparing as he advances in civilization, until the impoverishment, with which his exhaustion of the natural resources of the soil is threatening him, at last awakens him to the necessity of preserving what is left, if not of restoring what has been wantonly wasted.

🌿

From
"Wild Wool"
John Muir

No dogma taught by present civilization seems to form so insuperable an obstacle in the way of a right understanding of the relations which culture sustains to wildness as that which regards the world as made especially for the uses of man. Every animal, plant, and crystal controverts it in the plainest terms. Yet it is taught from century to century as something ever new and precious, and in the resulting darkness the enormous conceit is allowed to go unchallenged.

🌿

"Wilderness: A Human Need"
Sigurd F. Olson

For millions of years man shared ancient ecological balances and relationships with other creatures in a common environment. He is still therefore physiologically and spiritually part of a primeval past, still attuned to nature and never quite happy or content removed from its influences. Man became human because of the development of his brain, his ability to adapt himself to new situations, and the slow growth of awareness and perception. In spite of his sophistication and inventive genius, he still does not understand or appreciate the importance of natural beauty and of preserving some areas where the ancient scene is undisturbed.

Deep down in his subconscious, a part of his pool of racial memories is an abiding sense of oneness with life he cannot deny. Within him is a hunger and a craving for wildness and nature, which he cannot quite understand. He must feel the ground under his feet, use muscles as they were meant to be used, know the warmth and light of wood fires in primitive shelters away from storms. He must feel old rhythms, the cyclic change of seasons, see the miracles of growth, and sense the issues of life and death. He is, in spite of himself, still a creature of forests and open meadows, of rivers, lakes, and seashores. He needs to look at sunsets and sunrises and the coming of a full moon. Although he is conquering space and producing life, ancient needs and longings are still part of him, and in his urbanized technological civilization he still listens to the song of the wilderness.

The hidden forest in all its vagaries, its many facets, and its amazing interdependencies, is part of him. The forest is more than trees, soil, and water, far more than can ever be seen. It is a place where he can sense the world as it used to be, a sanctuary of the spirit where he can find himself in an eroding environment where the old values are being lost—the values that made him what he is.

In our swiftly growing civilization, with its new cities and communities and pyramiding industrial uses polluting the air, soil, and water and bringing ugliness to places of beauty, man's problems are very real. The vast concentrations of people in inhuman megalopolises, removed from a natural to an increasingly artificial environment, make preservation of wild or unblemished areas of inestimable value.

Such places should be reserved from development or exploitation for the simple reason that they are needed by modern man for solace and balance. No trees should be cut there, no machines allowed to scar the earth's surface, no pollution permitted, no wild creatures or vegetation eliminated. Disturb such places in the slightest, and they lose their value. In the face of the imbalance, man's spirit shrivels, for only in nature can he gain perspective.

It has been said that beauty is in the eye of the beholder and that it is its own excuse for being. Man needs beauty as he needs food. He is constantly in search of it. Artists spend their lives creating it, but for the vast majority it lies in the simplicities of natural things. Ugliness is revolting to man, but beauty sustains him. The hidden forest is vital to man's happiness, but only by being aware of the unseen forces at work within it can he truly appreciate its worth.

🎋

From
"Eco-Interests"
Lawrence E. Johnson

I claim that an ecosystem is the sort of thing that can have interests, that ecosystems do have interests, and that their interests are morally signifi-cant. In claiming that the woods as well as the individual trees are morally significant, I claim that an ecosystem is more than a collection of various living beings that have their own interests. Individual organisms are certainly involved, and their interests do count, but ecosystems have morally signif-icant interests that are not just the aggregated interests of the individual organisms. Ecosystems have, in a legitimate sense, a life of their own.

🎋

From
"The Wilderness
Idea Reaffirmed"
Holmes Rolston III

(Editor's note: This essay is in opposition to the opinions of J. Baird Callicott, who argues for "the sustainable development alternative" to wilderness.)

Introduction

Revisiting the wilderness, Callicott is a doubtful guide; indeed he has got-ten himself lost. That is a pity, because . . . I readily endorse his positive arguments for developing a culture more harmonious with nature. But these give no cause for being negative about wilderness.

The wilderness concept, we are told, is "inherently flawed," triply so. It metaphysically and unscientifically dichotomizes man and nature. It is ethnocentric, because it does not realize that practically all the world's ecosystems were modified by aboriginal peoples. It is static, ignoring change through time. In the flawed idea and ideal, wilderness respects wild communities where man is a visitor who does not remain. In the revisited idea, also Leopold's ideal, humans, themselves entirely natural, reside in and can and ought to improve wild nature.

Human Culture and Wild Nature

Wilderness valued without humans perpetuates a false dichotomy, Callicott maintains. Going back to Cartesian and Greek philosophy and Christian theology, such a contrast between humans and wild nature is a metaphysical confusion that leads us astray and also is unscientific. But this is not so. One hardly needs metaphysics or theology to realize that there are critical differences between wild nature and human culture. Humans now superimpose cultures on the wild nature out of which they once emerged. There is nothing unscientific or non-Darwinian about the claim that innovations in human culture make it radically different from wild nature.

Information in wild nature travels intergenerationally on genes; information in culture travels neurally as persons are educated into transmissible cultures. In nature, the coping skills are coded on chromosomes. In culture, the skills are coded in craftsman's traditions, religious rituals, or technology manuals. Information acquired during an organism's lifetime is not transmitted genetically; the essence of culture is acquired information transmitted to the next generation. Information transfer in culture can be several orders of magnitude faster and overleap genetic lines. I have but two children; copies of my books and my former students number in the thousands. A human being develops typically in some one of ten thousand cultures, each heritage historically conditioned, perpetuated by language, conventionally established, using symbols with locally effective meanings. Animals are what they are genetically, instinctively, environmentally, without any options at all. Humans have myriad lifestyle options, evidenced by their cultures; and each human makes daily decisions that affect his or her character. Little or nothing in wild nature approaches this.

The novelty is not simply that humans are more versatile in their spontaneous natural environments. Deliberately rebuilt environments replace spontaneous wild ones. Humans can therefore inhabit environments altogether different from the African savannas in which they once evolved. They insulate themselves from environmental extremes by their rebuilt habitations, with central heat from fossil fuel or by importing fresh groceries from a thousand miles away. In that sense, animals have freedom within ecosystems, but humans have freedom from ecosystems. Animals are adapted to their niches; humans adapt their ecosystems to their needs. The determinants of animal and plant behavior, much less the determinants of climate or nutrient recycling, are never anthropological, political, economic, technological, scientific, philosophical, ethical, or religious. Natural selection pressures are relaxed in culture; humans help each other out compassionately with medicine, charity, affirmative action, or Head Start programs.

Humans act using large numbers of tools and things made with tools, extrasomatic artifacts. In all but the most primitive cultures, humans teach each other how to make clothes, thresh wheat, make fires, bake bread. Animals do not hold elections and plan their environmental affairs; they do not make bulldozers to cut down tropical rainforests. They do not fund development projects through the World Bank or contribute to funds to save the whales. They do not teach their religion to their children. They do not write articles revisiting and reaffirming the idea of wilderness. They do not get confused about whether their actions are natural or argue about whether they can improve nature.

If there is any metaphysical confusion in this debate, we locate it in the claim that "man is a natural, a wild, an evolving species, not essentially different in this respect from all the others." Poets like Gary Snyder perhaps are entitled to poetic license. But philosophers are not, especially when analyzing the concept of wildness. They cannot say that "the works of man, however precocious, are as natural as those of beavers," being "entirely natural," and then, hardly taking a breath, say that "the cultural component in human behavior is so greatly developed as to have become more a difference of kind than of degree." If this were only poetic philosophy it might be harmless, but proposed as policy, environmental professionals who operate with such contradictory philosophy will fail tragically.

"Anthropogenic changes imposed upon ecosystems are as natural as any other." Not so. Wilderness advocates know better; they do not gloss over these differences. They appreciate and criticize human affairs, with insight into their radically different character. Accordingly, they insist that there are intrinsic wild values that are not human values. These ought to be preserved for whatever they can contribute to human values, and because they are valuable on their own, in and of themselves. Just because the human presence is so radically different, humans ought to draw back and let nature be. Humans can and should see outside their own sector, their species self-interest, and affirm nonanthropogenic, noncultural values. Only humans have conscience enough to do this. That is not confused metaphysical dichotomy; it is axiological truth. To think that human culture is nothing but a natural system is not discriminating enough. It risks reductionism and primitivism.

These contrasts between nature and culture were not always as bold as they now are. Once upon a time, culture evolved out of nature. The early hunter–gatherers had transmissible cultures but, sometimes, were not much different in their ecological effects from the wild predators and omnivores among whom they moved. In such cases, this was as much through lack of power to do otherwise as from conscious decision. A few such aboriginal peoples may remain.

But we Americans do not and cannot live in such a twilight society. Any society that we envision must be scientifically sophisticated, technologically advanced, globally oriented, as well as (we hope) just and charitable, caring for universal human rights and for biospheric values. This society will try to fit itself in intelligently with the ecosystemic processes on which it is superposed. It will, we plead, respect wildness. But none of these decisions shaping society are the processes of wild nature. There is no inherent flaw in our logic when we are discriminating about these radical discontinuities between culture and nature. The dichotomy charge is a half-truth, and, taken for the whole, becomes an untruth.

Humans Improving Wild Nature

Might a mature, humane civilization improve wild nature? Callicott thinks that it is a "fallacy" to think that "the best way to conserve nature is to protect it from human habitation and utilization." But, continuing the analysis, surely the fallacy is to think that a nature allegedly improved by humans is anymore real nature at all. The values intrinsic to wilderness cannot, on pain of both logical and empirical contradiction, be "improved" by deliberate human management, because deliberation is the antithesis of wildness. That is the sense in which civilization is the "antithesis" of wilderness, but there is nothing "amiss" in seeing an essential difference here. Animals take nature ready to hand, adapted to it by natural selection, fitted into their niches; humans rebuild their world through artifact and heritage, agriculture and culture, political and religious decisions.

On the meaning of "natural" at issue here, that of nature proceeding by evolutionary and ecological processes, any deliberated human agency, however well intended, is intention nevertheless and interrupts these spontaneous processes and is inevitably artificial, unnatural. The architectures of nature and of culture are different, and when culture seeks to improve nature, the management intent spoils the wildness. Wilderness management, in that sense, is a contradiction in terms—whatever may be added by way of management of humans who visit the wilderness, or of restorative practices, or monitoring, or other activities that environmental professionals must sometimes consider. A scientifically managed wilderness is conceptually as impossible as wildlife in a zoo.

To recommend that *Homo sapiens* "reestablish a positive symbiotic relationship with other species and a positive role in the unfolding of evolutionary processes" is, so far as wilderness preservation is involved, not just bad advice, it is impossible advice. The cultural processes by their very "nature" interrupt the evolutionary process; there is no symbiosis, there is antithesis. Culture is a post-evolutionary phase of our planetary history; it must be superposed on the nature it presupposes. To recommend, however, that we should build sustainable cultures that fit in with the continuing ecological processes is a first principle of intelligent action, and no wilderness advocate thinks otherwise.

If there are inherent conceptual flaws dogging this debate, we have located another: Callicott's allegedly "improved" nature. In such modified nature, the different historical genesis brings a radical change in value type. Every wilderness enthusiast knows the difference between a pine plantation in the Southeast and an old-growth grove in the Pacific Northwest. Even if the "improvement" is more or less harmonious with the ecosystem, it is fundamentally of a different order. Asian ring-tailed pheasants are rather well naturalized on the contemporary Iowa landscape. But they are there by human introduction, and they remain because farmers plow the fields, plant corn, and leave shelter in the fencerows. They are really as much pets as like native wild species because they are not really on their own. . . .

"The Saving Wildness"
Thomas J. Lyon

La Garita Wilderness, Colorado. August 4. I took off my pack, set it against a young Engelmann spruce, and had another look at the map: crossing Cochetopa Creek and then going steeply uphill through the green, gaining about fifteen hundred feet, I would come into a high, treeless alpine bowl, and above that cross an easy divide at around 13,000 feet.

It looked O.K. on paper, but the steep woods turned out to be barricaded with down trees, and despite the angle the footing was boggy much of the way. It was a struggle going up, longer than I'd thought—and, finally a great release, at the end of the woods, to step out onto firm ground. Now, moving in the easy, grass country above treeline, I looked across the canyon of Cochetopa to the broad, tilted uplands on Organ Mountain, and the triangular summit of San Luis Peak, and way out to the far, snowy peaks of the western San Juans . . . a huge view. The rising, jagged points of the earth glittered in the air. In high, open country like this, walking on the close-knit turf and jumping from rock to rock, the body takes on unexpected vitality, an energy that seems inexhaustible. You realize you're grinning. A wiggle of delight runs outward from an unknown source, involving every nerve. Who knows what makes the heart rise up? The joy of the body is the joy of the mind, too—feeling the mountains in your bones, their slope and bulk moving with you. They feel like your shoulders and knees somehow. "The blue mountains are constantly walking," said the Japanese philosopher Dogen.

On the broad divide the wind was blowing, as it always seems to in such bare terrain. Something made me walk even slower now, and notice better—see better. There was a big slanted rock, among the many sticking out of the turf at just the right angle; I let the pack down again and lay back against the warm rock. Speckled all around were the low flower-cushions of the tundra, clamped to the ground, their blossoms barely nudged by the wind that moved the taller grasses: little myriad white crosses of alpine phlox, pink moss campion, and blue forget-me-nots. The whole broad saddle sparkled with color in the airy light. I took off boots and socks and put my bare feet on the wiry grass. After a time the usual midday clouds began to build up. A cloud would come, and the light would shut down suddenly, and then after a minute or two suddenly come back. My eyes and ears seemed to widen, become more alert, at each change.

Logan, Utah. January 21. Now I'm at the office, typing on a word processor. A good soldier, I will go to a faculty meeting at 12:30. I am, of course, a working member of the civilization that is ruining the world. Why are we doing this?—so helplessly, so resistlessly? I have no great answer. But I wouldn't even ask the question, without the wilderness. I think back to that curving, soft-formed ridge, the look of it in the sliding shadow and light; the warm rocks, naked to the sky; the surge of plain good feeling, and under the feeling an inward awareness, a pure arc of knowing. The beauty of wilderness is the beauty of essence, of the whole. It is the source, where our life comes from, and the body knows this.

So it seems off the mark to argue for wilderness as we do, on the basis of watershed protection, or recreation, or big game habitat, or to save a plant that might one day give a cure for cancer. Not wrong, exactly, because all these points are unarguable, or so I believe—but they're not the real point. And it seems weird, from a holistic point of view, that the wilderness, the most shiningly clear example of the world's overall integrity, is fragmented in our law into managed bits and pieces. The Wilderness Act of 1964 says that wilderness, "in contrast with those areas where man and his own works dominate the landscape, is hereby recognized as an area where the earth and its community of life are untrammeled by man, where man himself is a visitor who does not remain." This is politically and historically realistic, given our particular culture, but in the long run such dualism goes against the grain of things.

The new-age truism that "health," "whole," and "holy" are a word-family really is true. Their connateness represents an ancient understanding that nothing separate can be genuinely sound. Health is cooperative, a matter of good citizenship in a system. There is no separate, cocooned salvation. "What is the use of a house," Henry David Thoreau asked, "if you haven't got a tolerable planet to put it on?" We begin slowly to recognize the togetherness of all things that the etymology of "heal" hints at. Aldo Leopold called the development of the ecological point of view the outstanding advance of the twentieth century. The real work, to borrow Gary Snyder's phrase, is to put that point of view into practice.

I think such a deep-seated revolution is possible, but only if it is felt to involve beauty, joy, and delight. I don't believe we can make a future out of technological fixes—we have bandaids on top of bandaids even now—and I don't imagine that guilt or authority are workable, long-term motivations. What we crave, what's missing, is physical and emotional experience of the wild Earth, the contact that sparks our oldest delight back into life. Only from hints of joyfulness and health can we be inspired at a deep enough level. Starting from anywhere else, we get merely consolation, or partial repair. The theme was sounded 130 years ago by Thoreau: "In Wildness is the preservation of the World."

At present, and it's well to acknowledge this, we all join the economy. Following the accepted strategy, we fill each day with things to do, ensuring that the small kind of consciousness will occupy the whole of waking existence. Every day, working together, we make more of the wild Earth our own—that is, break it into parts the ego can administer. As the managed world grows, the wild, mysterious sense of relation atrophies and our stock of metaphors, narrowed to our own creations, becomes pathetically self-referential. "A falcon is built for speed," says a special supplement to *Audubon* magazine, September, 1990. "Powerful but streamlined, a falcon can streak across the sky like a jet fighter, drop like a dive-bomber, turn on a dime."

This kind of thing forecasts death from solipsism, and shows why we now need wilderness so profoundly. When we walk for days in a natural area, and spend the nights there, the body begins to come alive in a way that is unmistakably different from what jogging or weight-lifting do. A certain competence and confidence begin to seem natural, as feet, working now in variable terrain, adjust themselves to rocks and ground. Hands start to feel wider and stronger. The body brings the mind along. The sense of place comes from the simple enjoyment of surroundings, meeting them, touching them and being touched. The redeeming wildness in us arises in practical response to actual landscapes. This is where natural equanimity lies—it is where we are not constitutionally needy.

Am I just thinking here about humans, and our own problem? I hope not, I think not. Our completion is now crucial for everything else in the world, because only by restoring our own health will we stop destroying the planet in the name of our so-called needs.

Must the setting for this redemption be wilderness? I believe so. If we stand on the tamed realm, all we see are signs and symbols of fragmentation. The managed part of the world corroborates reduced consciousness and the little, needy self: it shows us ourselves again and again, like a hall of mirrors. In the wild, such constant reinforcement simply isn't there. We are surrounded and moved now by spontaneous life—a flashing stream; a path obscured by fallen trees; the deep, enveloping sound of wind coming through a forest at night. Such things retain an immense, archaic potency with us. They touch and awaken our own share of the oldest wisdom, the accumulated intelligence of the Earth.

The "Late Summer, 1990" issue of *Aperture* magazine, titled *Beyond Wilderness,* warns against focusing on wilderness so strongly that we neglect our other environmental responsibilities. What we ought to be doing, write Barry Lopez and Wes Jackson for example, is seeing wilderness as just one aspect of the whole span of relations between humanity and nature. We need to be doing better by all ecosystems and every place we touch, not just romantically setting apart a few pristine wild areas and then going on with business as usual. Lopez concludes, "Wild landscapes are necessary to our being. We require them as we require air and water. But we need, at the same time, to create a landscape in which wilderness makes deep and eminent sense as a part of the whole, a landscape in which wilderness is not an orphan." Such a wise statement is hard to take exception to. If we followed it, we would have a better world than we have now. But having said that, I recommend a different slant; I would prefer a thought like Robert Frost's, expressed in a 1934 poem, "Unharvested."

> May something go always unharvested!
> May much stay out of our stated plan.

In our heart of hearts, wilderness isn't part of anything. It is the overarching reality that transcends all our plans and creations. We cannot go "beyond wilderness"—the universe is wild. We can only go beyond our paltry, dichotomized worldview and our management-mania. The wilderness that is left on this planet (one-third of the land area, according to the most hopeful estimate) is where we should look for our original reference and inspiration; it should not be considered just one of the many competing possibilities to which, in stewardship, we may parcel out the Earth. In the wilderness battle is involved the whole issue of how we conceive of ourselves and our powers. It is right, surely, to reintroduce peregrine falcons, try to breed perennial cereal crops, and make electricity by the sun. There is a tremendous lot we need to do, to be good citizens. But what we need most of all is a radical, inward change, a restoration of our perception and our joy in the world. What needs to come forth is our wildness, which is to say, our wholeness. Then we would have a standard that could give meaning and proportion to our amazing techniques.

"Anthropocentrism"
John Seed

But the time is not a strong prison either.
A little scraping the walls of dishonest contractor's concrete
Through a shower of chips and sand makes freedom.
Shake the dust from your hair. This mountain sea-coast is real
For it reaches out far into the past and future;
It is part of the great and timeless excellence of things.[1]

"Anthropocentrism" or "homocentrism" means human chauvinism. Similar to sexism, but substitute "human race" for "man" and "all other species" for "woman."

Human chauvinism, the idea that humans are the crown of creation, the source of all value, the measure of all things, is deeply embedded in our culture and consciousness.

And the fear of you and the dread of you shall be upon every beast of the earth, and upon every fowl of the air, and upon all that moveth on the earth, and upon all the fishes of the sea; into your hands they are delivered.[2]

When humans investigate and see through their layers of anthropocentric self-cherishing, a most profound change in consciousness begins to take place.

Alienation subsides. The human is no longer an outsider, apart. Your humanness is then recognized as being merely the most recent stage of your existence . . . you start to get in touch with yourself as mammal, as vertebrate, as a species only recently emerged from the rain forest. As the fog of amnesia disperses, there is a transformation in your relationship to other species, and in your commitment to them.

What is described here should not be seen as merely intellectual. The intellect is one entry point to the process outlined, and the easiest one to communicate. For some people, however, this change of perspective follows from actions on behalf of mother Earth.

"I am protecting the rain forest" develops to "I am part of the rain forest protecting myself. I am that part of the rain forest recently emerged into thinking."

What a relief then! The thousands of years of imagined separation are over and we begin to recall our true nature. That is, the change is a spiritual one, thinking like a mountain,[3] sometimes referred to as "deep ecology."

As your memory improves, as the implications of evolution and ecology are internalized and replace the outmoded anthropocentric structures in your mind, there is an identification with all life. Then follows the realization that the distinction between "life" and "lifeless" is a human construct. Every atom in this body existed before organic life emerged 4,000 million years ago. Remember our childhood as minerals, as lava, as rocks?

Rocks contain the potentiality to weave themselves into such stuff as this. We are rocks dancing. Why do we look down on them with such a condescending air? It is they that are the immortal part of us.[4]

If we embark upon such an inner voyage, we may find, upon returning to 1983 consensus reality, that our actions on behalf of the environment are purified and strengthened by the experience.

We have found here a level of our being that moth, rust, nuclear holocaust, or destruction of the rain forest gene pool do not corrupt. The commitment to save the world is not decreased by the new perspective, although the fear and anxiety which were part of our motivation start to dissipate and are replaced by a certain disinterestedness. We act because life is the only game in town, but actions from a disinterested, less attached consciousness may be more effective.

Activists often don't have much time for meditation. The disinterested space we find here may be similar to meditation. Some teachers of meditation are embracing deep ecology[5] and vice versa[6].

Of all the species that have ever existed, it is estimated that less than one in a hundred exist today. The rest are extinct.

As environment changes, any species that is unable to adapt, to change, to evolve, is extinguished. All evolution takes place in this fashion. In this way an oxygen-starved fish, ancestor of yours and mine, commenced to colonize the land. Threat of extinction is the potter's hand that moulds all the forms of life.

The human species is one of millions threatened by imminent extinction through nuclear war and other environmental changes. And while it is true that "human nature" revealed by 12,000 years of written history does not offer much hope that we can change our warlike, greedy, ignorant ways, the vastly longer fossil history assures us that we *can* change. We *are* that fish, and the myriad other death-defying feats of flexibility which a study of evolution reveals to us. A certain confidence (in spite of our recent "humanity") is warranted.

From this point of view, the threat of extinction appears as the invitation to change, to evolve. After a brief respite from the potter's hand, here we are back on the wheel again.

The change that is required of us is not some new resistance to radiation, but a change on consciousness. Deep ecology is the search for a viable consciousness.

Surely consciousness emerged and evolved according to the same laws as everything else—moulded by environmental pressures. In the recent past, when faced with intolerable environmental pressures, the mind of our ancestors must time and again have been forced to transcend itself.

To survive our current environmental pressures, we must consciously remember our evolutionary and ecological inheritance. We must learn to think like a mountain.

If we are to be open to evolving a new consciousness, we must fully face up to our impending extinction (the ultimate environmental pressure). This means acknowledging that part of us which shies away from the truth, hides in intoxication or busyness from the despair of the human, whose 4,000 million year race is run, whose organic life is a mere hair's breadth from finished.[7]

A biocentric perspective, the realization that rocks *will* dance, and that roots go deeper than 4,000 million years, may give us the courage to face despair and break through to a more viable consciousness, one that is sustainable and in harmony with life again.

> Protecting something as wide as this planet is still an abstraction for many. Yet I see the day in our lifetime that reverence for the natural systems—the oceans, the rain forests, the soil, the grasslands, and all other living things—will be so strong that no narrow ideology based upon politics or economics will overcome it. (Jerry Brown, Governor of California)[8]

NOTES

1. From the poem "A Little Scraping," *The Selected Poetry of Robinson Jeffers* (New York: Random House, 1933, out of print).

2. Genesis 9:2.
3. "The forester ecologist Aldo Leopold underwent a dramatic conversion from the 'stewardship' shallow ecology resource-management mentality of man-over-nature to announce that humans should see themselves as 'plain members' of the biotic community. After the conversion, Leopold saw steadily, and with 'shining clarity' as he broke through the anthropocentric illusions of his time and began 'thinking like a mountain.'" George Sessions, "Spinoza, Perennial Philosophy and Deep Ecology" (photostat, Sierra College, Rocklin, California, 1979). See also Aldo Leopold, *A Sand County Almanac* (London: O.U.P., 1949).
4. Prominent physicists such as David Bohm (*Wholeness and the Implicate Order,* Routledge, 1980), and biologists and philosophers such as Charles Birch and John Cobb, Jr. (*The Liberation of Life,* Cambridge, 1981) would agree with Alfred North Whitehead that "A thoroughgoing evolutionary philosophy is inconsistent with materialism. The aboriginal stuff, or material from which a materialistic philosophy starts, is incapable of evolution." (*Science and the Modern World,* Fontana, 1975 [first published 1926], p. 133). Similar views to those of these authors on the interpenetration of all "matter" (better conceived of as "events") are developed in Fritjof Capra's *The Tao of Physics* (Fontana, 1976), while the sixth-century B.C. *Tao Te Ching* itself tells us that "Tao" or "the implicate order," as Bohm might say, "is the source of the ten thousand things" (translated by G. Feng and J. English, New York: Vintage, 1972).
5. "For Dogen Zenji, the others who are 'none other than myself' include mountains, rivers, and the great earth. When one thinks like a mountain, one thinks also like the black bear, and this is a step . . . to deep ecology, which requires openness to the black bear, becoming truly intimate with the black bear, so that honey dribbles down your fur as you catch the bus to work." Robert Aitken Roshi, Zen Buddhist teacher, "Gandhi, Dogen and Deep Ecology," *Zero,* 4 (1980).
6. Theodore Roszak, for example, has written in *Person/Planet* (Victor Gollanz, 1979, p. 296): "I sometimes think there could be no keener criterion to measure our readiness for an economics of permanence than silence." Roszak has argued eloquently in another context that, if ecology is to work in the service of transforming consciousness, it will be because its students recognize the truth contained in a single line of poetry by Kathleen Raine: "It is not birds that speak, but men learn silence." (*Where the Wasteland Ends,* Massachusetts: Faber and Faber, 1974, p. 404).
7. For the creative uses of despair, see Joanna Macy, "Despair Work," *Evolutionary Blues* 1 (1981). For a long look at our impending extinction, see Jonathan Schell, *The Fate of the Earth* (Pan Books, 1982).
8. "Not Man Apart," Friends of the Earth newsletter, 9, 9 (August, 1979).

❧

"The Myths We Live By"
George Wuerthner

A few years ago, I spent an entire day in Yellowstone National Park with one of the most outspoken critics of the park's wildlife policies. He believed park officials were guilty of malfeasance for permitting elk and other wildlife to self-regulate their populations. Such a policy, he felt, was destroying the park's vegetation and jeopardized the park's landscapes. Although I had been to Yellowstone many times, my own observations didn't jive with his perspective. So, thinking that perhaps I was missing something, I asked him to spend a day with me in the park and make his best case, an invitation he eagerly accepted.

As we wandered the northern range looking at plants and talking about management policies, I began to learn his "story" of the role of people and Nature in the West. It became evident to me that the park's natural regulation policy, which he perceived as the elk "problem," actually represented a deeper, more fundamental challenge to his belief system. What he really didn't like about Yellowstone is that Nature appeared to be out of control—specifically, beyond human control.

He didn't like it that elk died from starvation or were killed by bears and wolves, and thus "were wasted"—unavailable to be taken by hunters and consumed by people. And to let "timber" (rather than trees) burn up in fires seemed to him to be the equivalent of a Holocaust against forests. As he explained to me while we were driving back to Bozeman, "God put those trees on the hillside for people to use and to just let them rot or burn up is going against His teachings."

His views on Nature, particularly their obvious religious foundation, may represent the extreme end of a particular perspective, but are not, at their heart, all that uncommon. Indeed, to one degree or another his perspective represents the dominant worldview of most natural resource managers, loggers, ranchers, miners, and other commercial users of the land. Their livelihoods depend upon control and manipulation of the land and its wildlife to meet human ends. All require a "domestication" of the landscape.

This is distinctly different from the goals of wildlands preservationists who seek to protect or restore self-generating and self-regulating landscapes and wildlife populations. Wild landscapes are those where human control and manipulation is minimal; as such, they threaten the values of those who seek to domesticate our forests and grasslands. These divergent views on how the world is ordered, and how humans fit into that world, are at the core of most environmental conflicts. The controversy over wolf restoration exemplifies the divergent parables. Anyone who sees this debate as solely about biology or economics misses a very important point: Ranchers and others who advocate human control of the landscape fear the wolf not only because it may occasionally consume one of their cows or sheep, but also because it represents a challenge to the dominant cultural myth of the Western Frontier—a bucolic agricultural landscape where livestock are tended by "hard-working" cowboys.

While conservationists may primarily base their advocacy of wolf recovery upon ecological arguments, for many, wolf restoration is also an attempt to "rewrite" the story of the West. It requires humans to give up willingly and freely a certain degree of control and manipulation of the land. Thus, wolf restoration is accurately viewed by wolf opponents as a direct attack upon the dominant parable that organizes their lives. The passions that lie behind the debate over wolf restoration are so fierce because they involve fundamental assumptions about the human–Nature compact.

Ironically, though this debate is primarily one about values, both sides of the wolf restoration debate (and most other natural resources conflicts) extensively rely upon scientific studies to bolster the legitimacy of their positions. Unfortunately, what is being debated cannot be answered by science. Decisions about wolf restoration, whether to graze cattle in the arid West, whether to kill bison outside of parks, whether to log forests to "save" them, or "protect" landscapes as "wilderness," and other current controversies are, at their roots, debates over the stories we want to tell ourselves. We may give science a holy place at the altar, but in reality, what guides our decisions and fuels our passions are the myths we live by. Science may be able to tell us that cows trampling a riparian zone results in fewer fish in our streams, or that logging old-growth forests causes spotted owls to go extinct, but whether that is perceived as a problem or not depends upon one's values—and these values are shaped by the stories we use to guide our lives.

Certainly, science is a powerful tool to help us see connections and relationships; but it is the vision and the way it is interpreted, not the science, that will capture people's hearts, and ultimately their minds. Because many environmental issues involve deeply held ideas about our perception of Nature and the human relationship to the natural world, the idea that science and rational debate can sway the outcome seems a bit optimistic, perhaps even naive.

Rather, it may be the poets, musicians, writers, and artists who will communicate a new vision of the American West as a place where people live among bison herds, where streams full of trout flow without being dewatered, and wolves are more than token animals in a few National Parks. It is the storytellers who ultimately may change the western parable, and thus, our relationship to the land and Nature.

From
Ways of Nature
John Burroughs

We marvel at what we call the wisdom of Nature, but how unlike our own! How blind, and yet in the end how sure! How wasteful, and yet how conserving! How helter-skelter she sows her seed, yet behold the forest or the flowery plain. Her springs leap out everywhere, yet how inevitably their waters find their way into streams, the streams into rivers, and the rivers to the sea. Nature is an engineer without science, and a builder without rules.

"Faking Nature"
Robert Elliot

Environmentalists express concern at the destruction/exploitation of areas of the natural environment because they believe that those areas are of intrinsic value. An emerging response is to argue that natural areas may have their value restored by means of the techniques of environmental engineering. It is then claimed that the concern of environmentalists is irrational, merely emotional or even straightforwardly selfish. This essay argues that there is a dimension of value attaching to the natural environment which cannot be restored no matter how technologically proficient environmental engineers become. The argument involves highlighting and discussing analogies between faking art and faking nature. The pivot of the argument is the claim that genesis is a determinant of an area's value.

Consider the following case. There is a proposal to mine beach sands for rutile. Large areas of dune are to be cleared of vegetation and the dunes themselves destroyed. It is agreed, by all parties concerned, that the dune area has value quite apart from a utilitarian one. It is agreed, in other words, that it would be a bad thing considered in itself for the dune area to be dramatically altered. Acknowledging this the mining company expresses its willingness, indeed its desire, to restore the dune area to its original condition after the minerals have been extracted.[1] The company goes on to argue that any loss of value is merely temporary and that full value will in fact be restored. In other words they are claiming that the destruction of what has value is compensated for by the later creation (recreation) of something of equal value. I shall call this "the restoration thesis."

In the actual world many such proposals are made, not because of shared conservationist principles, but as a way of undermining the arguments of conservationists. Such proposals are in fact effective in defeating environmentalist protest. They are also notoriously ineffective in putting right, or indeed even seeming to put right, the particular wrong that has been done to the environment. The sandmining case is just one of a number of similar cases involving such things as open-cut mining, clear-felling of forests, river diversion, and highway construction. Across a range of such cases some concession is made by way of acknowledging the value of pieces of landscape, rivers, forests, and so forth, and a suggestion is made that this value can be restored once the environmentally disruptive process has been completed.

Imagine, contrary to fact, that restoration projects are largely successful; that the environment is brought back to its original condition and that even a close inspection will fail to reveal that the area has been mined, clear-felled, or whatever. If this is so then there is temptation to think that one particular environmentalist objection is defeated. The issue is by no means merely academic. I have already claimed that restoration promises do in fact carry weight against environmental arguments. Thus Mr. Doug Anthony, the Australian Deputy Prime Minister, saw fit to suggest that sand-mining on Fraser Island could be resumed once the community becomes more informed and more enlightened as to what reclamation work is being carried out by mining companies. . . . [2] Or consider how the protests of environmentalists might be deflected in the light of the following report of environmental engineering in the United States.

> . . . about 2 km of creek 25 feet wide has been moved to accommodate a highway and in doing so engineers with the aid of landscape architects and biologists have rebuilt the creek to the same standard as before. Boulders, bends, irregularities, and natural vegetation have all been designed into the new section. In addition, special log structures have been built to improve the habitat as part of a fish development program. [3]

Not surprisingly the claim that revegetation, rehabilitation, and the like restore value has been strongly contested. J. G. Mosley reports that:

> The Fraser Island Environmental Inquiry Commissioners did in fact face up to the question of the relevance of successful rehabilitation to the decision on whether to ban exports (of beach sand minerals) and were quite unequivocal in saying that if the aim was to protect a natural area such success was irrelevant. . . . The inquiry said: " . . . even if, contrary to the overwhelming weight of evidence before the Commission, successful rehabilitation of the flora after mining is found to be ecologically possible on all mined sites on the Island . . . the overall impression of a wild, uncultivated island will be destroyed forever by mining." [4]

I want to show both that there is a rational, coherent ethical system which supports decisive objections to the restoration thesis, and that that system is not lacking in normative appeal. The system I have in mind will make valuation depend, in part, on the presence of properties which cannot survive the disruption–restoration process. There is, however, one point that needs clarifying before discussion proceeds. Establishing that restoration projects, even if empirically successful, do not fully restore value does not by any means constitute a knock-down argument against some environmentally disruptive policy. The value that would be lost if such a policy were implemented may be just one value among many which conflict in this situation. Countervailing considerations may be decisive and the policy thereby shown to be the right one. If my argument turns out to be correct it will provide an extra, though by no means decisive, reason for adopting certain environmentalist policies. It will show that the resistance which environmentalists display in the face of restoration promises is not merely silly, or emotional, or irrational. This is important because so much of the debate assumes that settling the dispute about what is ecologically possible automatically settles the value question. The thrust of much of the discussion is that if restoration is shown to be possible, and economically feasible, then recalcitrant environmentalists are behaving irrationally, being merely obstinate or being selfish.

There are indeed familiar ethical systems which will serve to explain what is wrong with the restoration thesis in a certain range of cases. Thus preference utilitarianism will support objections to some restoration proposal if that proposal fails to maximally satisfy preferences. Likewise classical utilitarianism will lend support to a conservationist stance provided that the restoration proposal fails to maximize happiness and pleasure. However, in both cases the support offered is contingent upon the way in which the preferences and utilities line up. And it is simply not clear that they line up in such a way that the conservationist position is even usually vindicated. While appeal to utilitarian considerations might be strategically useful in certain cases they do not reflect the underlying motivation of the conservationists. The conservationists seem committed to an account of what has value which allow that restoration proposals fail to compensate for environmental destruction despite the fact that such proposals would maximize utility. What then is this distinct source of value which motivates and underpins the stance taken by, among others, the Commissioners of the Fraser Island Environmental Inquiry?

IT IS INSTRUCTIVE TO LIST some reasons that might be given in support of the claim that something of value would be lost if a certain bit of the environment were destroyed. It may be that the area supports a diversity of plant and animal life, it may be that it is the habitat of some endangered species, it may be that it contains striking rock formations or particularly fine specimens of mountain ash. If it is only considerations such as these that contribute to the area's value then perhaps opposition to the environmentally disruptive project would be irrational provided certain firm guarantees were available; for instance that the mining company or timber company would carry out the restoration and that it would be successful. Presumably there are steps that could be taken to ensure the continuance of species diversity and the continued existence of the endangered species. Some of the other requirements might prove harder to meet, but in some sense or other it is possible to recreate the rock formations and to plant mountain ash that will turn out to be particularly fine specimens. If value consists of the presence of objects of these various kinds, independently of what explains their presence, then the restoration thesis would seem to hold. The environmentalist needs to appeal to some feature which cannot be replicated as a source of some part of a natural area's value.

Putting the point thus indicates the direction the environmentalist could take. He might suggest that an area is valuable partly because it is a natural area, one that has not been modified by human hand, one that is undeveloped, unspoilt, or even unsullied. This suggestion is in accordance with much environmentalist rhetoric, and something like it at least must be at the basis of resistance to restoration proposals. One way of teasing out the suggestion and giving it a normative basis is to take over a notion from aesthetics. Thus we might claim that what the environmental engineers are proposing is that we accept a fake or a forgery instead of the real thing. If the claim can be made good then perhaps an adequate response to restoration proposals is to point out that they merely fake nature; that they offer us something less than was taken away.[5] Certainly there is a weight of opinion to the effect that, in art at least, fakes lack a value possessed by the real thing.[6]

One way in which this argument might be nipped in the bud is by claiming that it is bound to exploit an ultimately unworkable distinction between what is natural and what is not. Admittedly the distinction between the natural and the non-natural requires detailed working out. This is something I do not propose doing. However, I do think the distinction can be made good in a way sufficient to the present need. For present purposes I shall take it that "natural" means something like "unmodified by human activity." Obviously some areas will be more natural than others according to the degree to which they have been shaped by human hand. Indeed most rural landscapes will, on this view, count as non-natural to a very high degree. Nor do I intend the natural/non-natural distinction to exactly parallel some dependent moral evaluations; that is, I do not want to be taken as claiming that what is natural is good and what is non-natural is not. The distinction between natural and non-natural connects with valuation in a much more subtle way than that. This is something to which I shall presently return. My claim then is that restoration policies do not always fully restore value because part of the reason that we value bits of the environment is because they are natural to a high degree. It is time to consider some counter-arguments.

An environmental engineer might urge that the exact similarity which holds between the original and the perfectly restored environment leaves no room for a value discrimination between them. He may urge that if they are *exactly* alike, down to minutest detail (and let us imagine for the sake of argument that this is a technological possibility), then they must be *equally* valuable. The suggestion is that value discriminations depend on there being intrinsic differences between the states of affairs evaluated. This begs the question against the environmentalist, since it simply discounts the possibility that events temporally and spatially outside the immediate landscape in question can serve as the basis of some valuation of it. It discounts the possibility that the manner of the landscape's genesis, for example, has a legitimate role in determining its value. Here are some examples which suggest that an object's origins do affect its value and our valuations of it.

Imagine that I have a piece of sculpture in my garden which is too fragile to be moved at all. For some reason it would suit the local council to lay sewerage pipes just where the sculpture happens to be. The council engineer informs me of this and explains that my sculpture will have to go. However, I need not despair because he promises to replace it with an exactly similar artifact, one which, he assures me, not even the very best experts could tell was not the original. The example may be unlikely, but it does have some point. While I may concede that the replica would be better than nothing at all (and I may not even concede that), it is utterly improbable that I would accept it as full compensation for the original. Nor is my reluctance entirely explained by the monetary value of the original work. My reluctance springs from the fact that I value the original as an aesthetic object, as an object with a specific genesis and history.

Alternatively, imagine I have been promised a Vermeer for my birthday. The day arrives and I am given a painting which looks just like a Vermeer. I am understandably pleased. However, my pleasure does not last for long. I am told that the painting I am holding is not a Vermeer but instead an exact replica of one previously destroyed. Any attempt to allay my disappointment by insisting that there just is no difference between the replica and the original misses the mark completely. There is a difference and it is one which affects my perception, and consequent valuation, of the painting. The difference of course lies in the painting's genesis.

I shall offer one last example which perhaps bears even more closely on the environmental issue. I am given a rather beautiful, delicately constructed, object. It is something I treasure and admire, something in which I find considerable aesthetic value. Everything is fine until I discover certain facts about its origin. I discover that it is carved out of the bone of someone killed especially for that purpose. This discovery affects me deeply and I cease to value the object in the way that I once did. I regard it as in some sense sullied, spoilt by the facts of its origin. The object itself has not changed but my perceptions of it have. I now know that it is not quite the kind of thing I thought it was, and that my prior valuation of it was mistaken. The discovery is like the discovery that a painting one believed to be an original is in fact a forgery. The discovery about the object's origin changes the valuation made of it, since it reveals that the object is not of the kind that I value.

What these examples suggest is that there is at least a prima facie case for partially explaining the value of objects in terms of their origins, in terms of the kinds of processes that brought them into being. It is easy to find evidence in the writings of people who have valued nature that things extrinsic to the present, immediate environment determine valuations of it. John Muir's remarks about Hetch Hetchy Valley are a case in point.[7] Muir regarded the valley as a place where he could have direct contact with primeval nature; he valued it, not just because it was a place of great beauty, but because it was also a part of the world that had not been shaped by human hand. Muir's valuation was conditional upon certain facts about the valley's genesis; his valuation was of a, literally, natural object, of an object with a special kind of continuity with the past. The news that it was a carefully contrived elaborate *ecological* artifact would have transformed that valuation immediately and radically.

The appeal that many find in areas of wilderness, in natural forests, and wild rivers depends very much on the naturalness of such places. There may be similarities between the experience one has when confronted with the multi-faceted complexity, the magnitude, the awesomeness of a very large city, and the experience one has walking through a rain forest. There may be similarities between the feeling one has listening to the roar of water over the spillway of a dam, and the feeling one has listening to a similar roar as a wild river tumbles down rapids. Despite the similarities there are also differences. We value the forest and river in part because they are representative of the world outside our dominion, because their existence is independent of us. We may value the city and the dam because of what they represent of human achievement. Pointing out the differences is not necessarily to denigrate either. However, there will be cases where we rightly judge that it is better to have the natural object than it is to have the artifact.

It is appropriate to return to a point mentioned earlier concerning the relationship between the natural and the valuable. It will not do to argue that what is natural is necessarily of value. The environmentalist can comfortably concede this point. He is not claiming that all natural phenomena have value in virtue of being natural. Sickness and disease are natural in a straightforward sense and are certainly not good. Natural phenomena such as fires, hurricanes, and volcanic eruptions can totally alter landscapes and alter them for the worse. All of this can be conceded. What the environmentalist wants to claim is that, within certain constraints, the naturalness of a landscape is a reason for preserving it, a determinant of its value. Artificially transforming an utterly barren, ecologically bankrupt landscape into something richer and more subtle may be a good thing. That is a view quite compatible with the belief that replacing a rich natural environment with a rich artificial one is a bad thing. What the environmentalist insists on is that naturalness is one factor in determining the value of pieces of the environment. But that, as I have tried to suggest, is no news. The castle by the Scottish loch is a very different kind of object, valuewise, from the exact replica in the appropriately shaped environment of some Disneyland of the future. The barrenness of some Cycladic island would stand in a different, better perspective if it were not brought about by human intervention.

As I have glossed it, the environmentalist's complaint concerning restoration proposals is that nature is not replaceable without depreciation in one aspect of its value which has to do with its genesis, its history. Given this, an opponent might be tempted to argue that there is no longer any such thing as "natural" wilderness, since the preservation of those bits of it which remain is achievable only by deliberate policy. The idea is that by placing boundaries around national parks, by actively discouraging grazing, trail-biking and the like, by prohibiting sand-mining, we are turning the wilderness into an artifact, that in some negative or indirect way we are creating an environment. There is some truth in this suggestion. In fact we need to take notice of it if we do value wilderness, since positive policies *are* required to preserve it. But as an argument against my over-all claim it fails. What is significant about wilderness is its causal continuity with the past. That is something that is not destroyed by demarcating an area and declaring it a national park. There is a distinction between the "naturalness" of the wilderness itself and the means used to maintain and protect it. What remains within the park boundaries is, as it were, the real thing. The environmentalist may regret that such positive policy is required to preserve the wilderness against human, or even natural, assault.[8] However, the regret does not follow from the belief that what remains is of depreciated value. There is a significant difference between preventing damage and repairing damage once it is done. That is the difference that leaves room for an argument in favor of a preservation policy over and above a restoration policy.

There is another important issue which needs highlighting. It might be thought that naturalness only matters insofar as it is perceived. In other words it might be thought that if the environmental engineer could perform the restoration quickly and secretly, then there would be no room for complaint. Of course, in one sense there would not be, since the knowledge which would motivate complaint would be missing. What this shows is that there can be loss of value without the loss being perceived. It allows room for valuations to be mistaken because of ignorance concerning relevant facts. Thus my Vermeer can be removed and secretly replaced with the perfect replica. I have lost something of value without knowing that I have. This is possible because it is not simply the states of mind engendered by looking at the painting, by gloatingly contemplating my possession of it, by giving myself over to aesthetic pleasure, and so on which explain why it has value. It has value because of the kind of thing that it is, and one thing that it is is a painting executed by a man with certain intentions, at a certain stage of his artistic development, living in a certain aesthetic *milieu*. Similarly, it is not just those things which make me feel the joy that wilderness makes me feel, that I value. That would be a reason for desiring such things, but that is a distinct consideration. I value the forest because it is of a specific kind, because there is a certain kind of causal history which explains its existence. Of course I can be deceived into thinking that a piece of landscape has that kind of history, has developed in the appropriate way. The success of the deception does not elevate the restored landscape to the level of the original, no more than the success of the deception in the previous example confers on the fake the value of a real Vermeer. What has value in both cases are objects which are of the kind that I value, not merely objects which I think are of that kind. This point, it should be noted, is appropriate independently of views concerning the subjectivity or objectivity of value.

An example might bring the point home. Imagine that John is someone who values wilderness. John may find himself in one of the following situations:

1. He falls into the clutches of a utilitarian-minded super-technologist. John's captor has erected a rather incredible device which he calls an experiencemachine. Once the electrodes are attached and the right buttons pressed one can be brought to experience anything whatsoever. John is plugged into the machine, and, since his captor knows full well John's love of wilderness, given an extended experience as of hiking through a spectacular wilderness. This is environmental engineering at its most extreme. Quite assuredly John is being short-changed. John wants there to be wilderness and he wants to experience it. He wants the world to be a certain way and he wants to have experiences of a certain king; veridical.

2. John is abducted, blindfolded, and taken to a simulated, plastic wilderness area. When the blindfold is removed John is thrilled by what he sees around him: the tall gums, the wattles, the lichen on the rocks. At least that is what he thinks is there. We know better: we know that John is deceived, that he is once again being short-changed. He has been presented with an environment which he thinks is of value but isn't. If he knew that the leaves through which the artificially generated breeze now stirred were synthetic he would be profoundly disappointed, perhaps even disgusted at what at best is a cruel joke.

3. John is taken to a place which was once devastated by strip-mining. The forest which had stood there for some thousands of years had been felled and the earth torn up, and the animals either killed or driven from their habitat. Times have changed, however, and the area has been restored. Trees of the species which grew there before the devastation grow there again, and the animal species have returned. John knows nothing of this and thinks he is in pristine forest. Once again, he has been short-changed, presented with less than what he values most.

In the same way that the plastic trees may be thought a (minimal) improvement on the experience machine, so too the real trees are an improvement on the plastic ones. In fact in the third situation there is incomparably more of value than in the second, but there could be more. The forest, though real, is not genuinely what John wants it to be. If it were not the product of contrivance he would value it more. It is a product of contrivance. Even in the situation where the devastated area regenerates rather than is restored, it is possible to understand and sympathize with John's claim that the environment does not have the fullest possible value. Admittedly in this case there is not so much room for that claim, since the environment has regenerated of its own accord. Still the regenerated environment does not have the right kind of continuity with the forest that stood there initially; that continuity has been interfered with by the earlier devastation. (In actual fact the regenerated forest is likely to be perceived quite different to the kind of thing originally there.)

I HAVE ARGUED that the causal genesis of forests, rivers, lakes, and so on is important in establishing their value. I have also tried to give an indication of why this is. In the course of my argument I drew various analogies, implicit rather than explicit, between faking art and faking nature. This should not be taken to suggest, however, that the concepts of aesthetic evaluation and judgment are to be carried straight over to evaluations of, and judgments about, the natural environment. Indeed there is good reason to believe that this cannot be done. For one thing an apparently integral part of aesthetic evaluation depends on viewing the aesthetic object as an intentional object, as an artifact, as something that is shaped by the purposes and designs of its author. Evaluating works of art involves explaining them, and judging them, in terms of their author's intentions; it involves placing them within the author's corpus of work; it involves locating them in some tradition and in some special *milieu*. Nature is not a work of art though works of art (in some suitably broad sense) may look very much like natural objects.

None of this is to deny that certain concepts which are frequently deployed in aesthetic evaluation cannot usefully and legitimately be deployed in evaluations of the environment. We admire the intricacy and delicacy of coloring in paintings as we might admire the intricate and delicate shadings in a eucalypt forest. We admire the solid grandeur of a building as we might admire the solidity and grandeur of a massive rock outcrop. And of course the ubiquitous notion of *the beautiful* has a purchase in environmental evaluations as it does in aesthetic evaluations. Even granted all this there are various arguments which might be developed to drive a wedge between the two kinds of evaluation, which would weaken the analogies between faking art and faking nature. One such argument turns on the claim that aesthetic evaluation has, as a central component, a judgmental factor, concerning the author's intentions and the like in the way that was sketched above.[9] The idea is that nature, like works of art, may elicit any of a range of emotional responses in viewers. We may be awed by a mountain, soothed by the sound of water over rocks, excited by the power of a waterfall, and so on. However, the judgmental element in aesthetic evaluation serves to differentiate it from environmental evaluation and serves to explain, or so the argument would go, exactly what it is about fakes and forgeries in art which discounts their value with respect to the original. The claim is that if there is no judgmental element in environmental evaluation, then there is no rational basis to preferring real to faked nature when the latter is a good replica. The argument can, I think, be met.

Meeting the argument does not require arguing that responses to nature count as aesthetic responses. I agree that they are not. Nevertheless there are analogies which go beyond emotional content, and which may persuade us to take more seriously the claim that faked nature is inferior. It is important to make the point that only in fanciful situations dreamt up by philosophers are there no detectable differences between fakes and originals both in the case of artifacts and in the case of natural objects. By taking a realistic example where there are discernible, and possibly discernible, differences between the fake and the real thing, it is possible to bring out the judgmental element in responses to, and evaluations of, the environment. Right now I may not be able to tell a real Vermeer from a Van Meegaran, though I might learn to do so. By the same token I might not be able to tell apart a naturally evolved stand of mountain ash from one which has been planted, but might later acquire the ability to make the requisite judgment. Perhaps an anecdote is appropriate here. There is a particular stand of mountain ash that I had long admired. The trees were straight and tall, of uniform stature, neither densely packed nor too open-spaced. I then discovered what would have been obvious to a more expert eye, namely that the stand of mountain ash had been planted to replace original forest which had been burnt out. This explained the uniformity in size, the density and so on: it also changed my attitude to that piece of landscape. The evaluation that I make now of that landscape is to a certain extent informed, the response is not merely emotive but cognitive as well. The evaluation is informed and directed by my beliefs about the forest, the type of forest it is, its condition as a member of that kind, its causal genesis, and so on. What is more, the judgmental element affects the emotive one. Knowing that the forest is not naturally evolved forest causes me to feel differently about it: it causes me to perceive the forest differently and to assign it less value than naturally evolved forests.

Val Routley has eloquently reminded us that people who value wilderness do not do so merely because they like to soak up pretty scenery.[10] They see much more and value much more than this. What they do see, and what they value, is very much a function of the degree to which they understand the ecological mechanisms which maintain the landscape and which determine that it appears the way it does. Similarly, knowledge of art history, of painting techniques, and the like will inform aesthetic evaluations and alter aesthetic perceptions. Knowledge of this kind is capable of transforming a hitherto uninteresting landscape into one that is compelling. Holmes Rolston has discussed at length the way in which an understanding and appreciation of ecology generates new values.[11] He does not claim that ecology reveals values previously unnoticed, but rather that the understanding of the complexity, diversity, and integration of the natural world which ecology affords us, opens up a new area of valuation. As the facts are uncovered, the values are generated. What the remarks of Routley and Rolston highlight is the judgmental factor which is present in environmental appraisal. Understanding and evaluation do go hand in hand; and the responses individuals have to forests, wild rivers, and the like are not merely raw, emotional responses.

NOT ALL FORESTS ARE ALIKE, not all rain forests are alike. There are countless possible discriminations that the informed observer may make. Comparative judgments between areas of the natural environment are possible with regard to ecological richness, stage of development, stability, peculiar local circumstance, and the like. Judgments of this kind will very often underlie hierarchical orderings of environments in terms of their intrinsic worth. Appeal to judgments of this kind will frequently strengthen the case for preserving some bit of the environment. . . .

One reason that a faked forest is not just as good as a naturally evolved forest is that there is always the possibility that the trained eye will tell the difference.[12] It takes some time to discriminate areas of Alpine plain which are naturally clear of snow gums from those that have been cleared. It takes some time to discriminate regrowth forest which has been logged from forest which has not been touched. These are discriminations which it is possible to make and which are made. Moreover, they are discriminations which affect valuations. The reasons why the "faked" forest counts for less, more often than not, than the real thing are similar to the reasons why faked works of art count for less than the real thing.

Origin is important as an integral part of the evaluation process. It is important because our beliefs about it determine the valuations we make.

It is also important in that the discovery that something has an origin quite different to the origin we initially believe that it has, can literally alter the way we perceive that thing.[13] The point concerning the possibility of detecting fakes is important in that it stresses just how much detail must be written into the claim that environmental engineers can replicate nature. Even if environmental engineering could achieve such exactitude, there is, I suggest, no compelling reason for accepting the restoration thesis. It is worth stressing though that, as a matter of strategy, environmentalists must argue the empirical inadequacy of restoration proposals. This is the strongest argument against restoration ploys, because it appeals to diverse value-frameworks, and because such proposals are promises to deliver a specific good. Showing that the good won't be delivered is thus a useful move to make.

NOTES

1. In this case *full* restoration will be literally impossible because the minerals are not going to be replaced.
2. J.B. Mosley, "The Revegetation 'Debate': A Trap For Conservationists," *Australian Conservation Newsletter,* Vol. 12 (1980), No. 8, p.1.
3. Peter Dunk, "How New Engineering Can Work with the Environment," *Habitat Australia,* Vol. 7 (1979), No. 5, p.12.
4. See Mosley, op. cit., p.1.
5. Offering something less is not, of course, always the same as offering nothing. If diversity of animal and plant life, stability of complex ecosystems, tall trees, and so on are things that we value in themselves, then certainly we are offered something. I am not denying this, and I doubt that many would qualify their valuations of the above-mentioned items in a way that leaves the restored environment devoid of value. Environmentalists would count as of worth programs designed to render polluted rivers reinhabitable by fish species. The point is rather that they may, as I hope to show, rationally deem it less valuable than what was originally there.
6. See, e.g., Colin Radford, "Fakes," *Mind,* Vol. 87 (1978), No. 345, pp. 66–76. and Nelson Goodman, *Languages of Art,* Bobbs-Merrill, New York, 1968, pp. 99–122, though Radford and Goodman have different accounts of why genesis matters.
7. See Ch. 10 of Roderick Nash, *Wilderness and the American Mind,* Yale University Press, New Haven, 1973.
8. For example, protecting the Great Barrier Reef from damage by the crown-of-thorns starfish.
9. See, e.g., Don Mannison, "A Prolegomenon to a Human Chauvinist

Aesthetic," in D.S. Mannison, M.A. McRobbie, R. Routley (Eds.), *Environmental Philosophy*, Research School of Social Sciences, Australian National University, Canberra, 1980, pp. 212–16.

10. Val Routley, "Critical Notice of Passmore's *Man's Responsibility for Nature*," *Australasian Journal of Philosophy*, Vol. 53 (1975), No.2, pp. 171–85.

11. Holmes Rolston III, "Is There An Ecological Ethic," *Ethics*, Vol. 85 (1975), No. 2, pp. 93–109.

12. For a discussion of this point with respect to art forgeries, see Goodman, op. cit., esp. pp. 103–12.

13. For an excellent discussion of this same point with respect to artifacts, see Radford, op.cit., esp. pp. 73–76.

"A Platform of the Deep Ecology Movement"

Arne Naess

(1) The flourishing of human and non-human life on Earth has intrinsic value. The value of non-human life forms is independent of the usefulness these may have for narrow human purposes.

(2) Richness and diversity of life forms are values in themselves and contribute to the flourishing of human and non-human life on Earth.

(3) Humans have no right to reduce this richness and diversity except to satisfy vital needs.

(4) Present human interference with the non-human world is excessive, and the situation is rapidly worsening.

(5) The flourishing of human life and cultures is compatible with a substantial decrease of the human population. The flourishing of non-human life requires such a decrease.

(6) Significant change of life conditions for the better requires change in policies. These affect basic economic, technological, and ideological structures.

(7) The ideological change is mainly that of appreciating *life quality* (dwelling in situations of intrinsic value) rather than adhering to a

high standard of living. There will be a profound awareness of the difference between big and great.

(8) Those who subscribe to the foregoing points have an obligation directly or indirectly to participate in the attempt to implement the necessary changes.

🦌

"The Grace of the Wild"
LAUDS
A Psalm of Praise: Wilderness Travel as an Artistic Act
Paul Gruchow

Being on the move, seeing what you have never seen before, not knowing where you will rest your head when night falls, receiving what comes as it comes, expecting everything and nothing: this is the allure of the canoe country. Every stroke of the paddle or step along the trail with a canoe on your shoulders or a pack on your back literally enlarges your world.

You travel under your own power and with the aid of ancient and elegantly simple tools—the canoe, the paddle, a pair of boots. Someone was traveling this way ten thousand years ago, and someone may be doing the same ten thousand years from now. In a world where change seems the only constant, where the past is increasingly suspect and the future ever more doubtful, it is exhilarating to be in touch with something that "binds together all humanity—the dead to the living and the living to the unborn."

The words are Joseph Conrad's, defining the work of the artist. Every earnest journey into the heart of the canoe country is, on the same terms, potentially a work of art, accessible to all.

The thinker, Conrad says, makes an appeal from ideas, and the scientist from facts. But "the artist appeals to that part of our being which is not dependent upon wisdom: to that in us which is a gift and not an acquisition—and, therefore, more permanently enduring. He speaks to our capacity for delight and wonder, to the sense of mystery surrounding our lives; to our sense of pity, and beauty, and pain; to the latent feeling of fellowship with all creation—and to the subtle but invincible conviction of solidarity that knits together the loneliness of innumerable hearts, to the solidarity in dreams, in joy, in sorrow, in aspirations, in illusions, in hope, in fear, which binds men to each other, which binds together all humanity—the dead to the living and the living to the unborn."

A wilderness journey makes just this appeal. It speaks:

To that part of our being which is not dependent upon wisdom: Not to what we have learned from books, but to whatever depends upon experience with the physical world: knowing how to read a footprint in the mud, how to steer a canoe into the wind, how to make a fire in the rain, what sort of weather the clouds and the wind foretell, where to look for a moose, which of the mushrooms in the forest are edible, whether the sound in the night is sinister or benign, which bird sings overhead, what flower blooms in the marsh.

To that which is a gift and not an acquisition: For nothing one encounters along the way can be possessed beyond the span of a single lifetime, or turned to any monetary advantage that is not destructive, or preserved unchanged for so much as twenty-four hours. Nothing here—not the mists that rise in the morning, nor the wind that blows at midday, nor the curtain of colors that falls in the evening, nor the slap of a beaver's tail in the night—can be commandeered, or caused to happen, or forbidden.

To our capacity for delight and wonder: The water tumbling from one lake into the next, the laughter of loons, the howling of wolves, the waft of cool air upon a sweaty brow, the silence in cedar swamps, the majesty of tall pines, the soaring of eagles, the sudden shimmering of lights in the northern sky at midnight: all these, freely given, daily remind us along the way of the grace abounding in the world.

To the sense of mystery surrounding our lives: The forest floor strewn with bones announces death; the string of bejeweled butterfly eggs laid out upon a leaf, the one mystery as large as God, life itself; and the beating of our own hearts, suddenly audible in the stillness of a moonless wilderness night, the thinness of the thread that binds the one to the other.

To our sense of pity, and beauty, and pain: Such old-fashioned words; pity—friendly sympathy, the desire to deliver mercy—being the most disreputable of the lot: "The gilded sheath of pity conceals the dagger of envy," Nietzsche, Conrad's contemporary, said, expressing the more modern view; and we have, perhaps, forgotten the connection between beauty and pain—the words *awful* and *awesome* come from the same root—until we have all day long battled a fierce wind blowing out of the west and, at last, with aching muscles, have made camp in some sheltered cove, discovering then the bliss that descends with the silence after the last light.

To the latent feeling of fellowship with all creation: the sudden conviction, arriving in quiet moments, that the pine dancing in the breeze, and the otter cavorting in the lake, and the loon laughing in the unseen distance are not aliens after all, but neighbors, distant, yet potentially knowable, like the stranger who rides the same bus as you every morning: Latent because it is, for most of us, a feeling poorly exercised. It comes alive, the Zen master Dogen reminds us, not because of what we ourselves experience, but because of all that is experienced in our presence. "To carry yourself forward and experience myriad things," he said, "is delusion. But myriad things coming forth and experiencing *themselves* is awakening."

The journey is indistinct from the traveler. As it is the instrument of awakening, so the traveler is aroused; what stirs is the inner voice of the artist.

🐾

The Imagination of the Earth
David Spangler

When the sun rises on the morning of January 1, 2000, it will undoubtedly be a dawn like any other in the Earth's four-and-a-half-billion year history. Nowhere is it recorded in the stones of the Earth, the currents of the ocean, or the tempests of the winds that that dawn shall herald some special event.

For much of humanity, however, that dawn will be special, marking the beginning of a new millennium. This passage into the twenty-first century is a social invention, an act of cultural imagination. As such, it provides a catalyst for other acts of imagination on which our collective prosperity and health—perhaps even our survival—may depend. It provides a context to reexamine, reevaluate, and where necessary, change attitudes and habits of culture that increasingly prove dysfunctional on the interconnected planet we inhabit.

One of these acts of reimagination concerns our relationship to the world. For nearly three centuries, Western culture has progressively imagined itself as distinct and separate from the natural order, while concomitantly imagining that order as simply a dead, material resource to be used

(and used up) according to the will of those governments and corporations capable of managing such exploitation. At the brink of the new millennium, there is a question whether this image can continue or whether
it spells disaster for all of us unless changed.

I take the position that we must reimagine the world and ourselves
within it in ways that recognize and emphasize the interconnectedness
and interdependency that exist within the natural order, an order that
includes humanity and all of its works.

One key area for this reimagination is that of western spirituality, theology, and ethics. The challenge is to find images (and a willingness to
embrace such images) that extend our context of spirituality to include
the nonhuman dimension of the world. We require an ecospirituality and
a bioethics that is not wholly anthropomorphic. We need to reimagine
Earth, spirit, and ourselves in ways that synthesize these three into a new
wholeness that is healing and empowering. In the words of ecotheologian
Thomas Berry, we seek a "new story" about ourselves and our world.

This is vital work, but it is important to remember that the objective is
not simply new images but a change of behavior. We need to embody a
new way of living and express a new generosity and expansiveness of
spirit. Images and myths are powerful tools that can assist this process, but
they can also turn in our hands and become obstructions. They can easily
and subtly shift from empowering and liberating inspirations into beliefs
and dogmas that substitute one orthodoxy for another. Having a new way
of thinking or believing about nature and ourselves, for instance, is not
the same thing as actually living out, through deep understanding and
attunement, a new holistic spirit. So the craft of reimagining the Earth
and spirit is a delicate one because we can short-circuit the process by
assuming too quickly that we have reached our goal.

To illustrate my point, I want to examine the imagination of the Earth
that flows from the scientific theory that our planet is itself a living organism, a theory known as the Gaia Hypothesis. In the decade or so since this
hypothesis was proposed, it has become a powerful image that has been
widely circulated, particularly in religious, ecological, and spiritual circles.
In some ways, it has become a symbol for the ecological and holistic sen-
sibility which for many people represent the essential qualities needed for
human survival. By examining this image and its implications, I can illu-
mine both the pitfalls and advantages involved in reimagining both Earth
and spirit.

In 1979, James Lovelock, a British atmospheric chemist and inventor,
published a book called *Gaia: A New Look at Life on Earth*. In it he pre-

sented a theory called the Gaia Hypothesis that he had developed in collaboration with American microbiologist, Lynn Margulis. This theory basically stated that the Earth's climate and surface environment are controlled by the planet's biosphere in a manner suggestive of the homeostatic self-regulation found in living organisms. Consequently, there is evidence that the planet functions as if it were a single living entity. At the suggestion of British novelist William Golding, a neighbor and friend of Lovelock, this planet-sized organism was named *Gaia* after the Greek goddess of the Earth.

Had Lovelock called his theory something less imaginative and more scientific, like the Theory of Environmental Regulation through Biospheric Homeostasis, our story would end here. The idea would have wound its way through the usual series of scientific publications and conferences, wrapped in mathematics and chemical jargon and would probably never have emerged into the public awareness enough to become a catchword for a new vision of the future. Cybernetic feedback loops are simply not the kind of images capable of firing the imagination and launching revolutions; it is difficult to imagine impassioned citizens picketing the White House or the headquarters of major corporations with banners proclaiming "Honor Biospheric Homeostasis or Else!"

Gaia, on the other hand, is exactly that kind of imaginative, inspiring image. It is the picture of an ancient goddess, arriving on the scene exactly as the feminist movement has been challenging the patriarchal view of the world. It resonates with ancient cosmologies held in nearly every culture on Earth at one time or another that affirmed that the Earth is a living being possessed of soul and consciousness. It is an image of mother, of home, of life, and of passion. It therefore touches us at a deep level of racial myth and memory. Furthermore, it wraps all this up in the currently acceptable form of a scientific image: now it is scientists, not just shamans, mystics, priestesses, and priests, who are telling us that the Earth is alive. Therefore it must be true! No wonder that in only two or three years after its original proposition, this idea had been taken up by religious, spiritual, feminist, and ecological groups around the world, who deluged the astonished Lovelock with hundreds of lecture invitations.

This is an amazing transposition of an idea from one realm into another. Gaia as a mythic idea is definitely alien to the original Gaia Hypothesis. Though the latter does conceive of the Earth as a living entity, such a being, in the words of Margulis, if conscious at all, has the sentiency "of an amoeba." Hardly the stuff of myth and spiritual invocation. But given the need of our time for new stories and new myths to help us nav-

igate difficult transitional crises, and subjected to the forces of reimagination now at work in our culture to provide such new myths, this hypothesis could hardly remain within the more sterile boundaries of "good" science. It is simply too fertile and powerful an image. So it has transcended
its origins to become a one-word hieroglyph suggesting a whole way of
life, an organizing principle of an entire new cultural outlook. In fact, as
suggested by such book titles as *Gaia: A New Way of Knowing,* a compila-
tion of essays edited by cultural historian William Irwin Thompson, Gaia
has come to signify a whole epistemology based on a holistic or systems
view of the world. (Of course, the emergence of this new epistemology
and cultural paradigm had been occurring anyway throughout the seventies and eighties; the image of Gaia did not bring it into being. It did, however, give it a powerful hook on which to hang the whole enterprise: a single imaginative word whose mythic resonances enabled it to symbolize
and summarize a nascent cultural movement already seeking coherence.)

Gaia has also taken on spiritual implications. After all, it was originally
the name of a goddess. While in its modern usage it is generally not used
to refer to or to reinvoke that ancient Grecian deity, it is often used to suggest the existence of a world soul or spirit. I have attended Christian worship services where in a spirit of progressiveness and ecological correctness, the participants call upon the "spirit of Gaia" to heighten their
awareness of their connections with the Earth. In this context, Gaia seems
to refer to a kind of oversoul or purposeful planetary spiritual presence
capable of communicating with or inspiring human consciousness, perhaps like an archangel presiding over the well-being of the world, or, in
psychological terms, perhaps like a collective awareness arising from all
the lives, human and nonhuman, that make up the Earth. At the very least,
the existence of such a world soul requires us to reexamine our modern
relationship with the planet, ceasing to see it as simply a resource of "dead
matter" but instead viewing it as a fellow being with whom we have some
form of responsible relationship. (Lovelock himself speculated that
humanity might be the nervous system of such a planetary being—the
emergence within that being of a capacity to think and imagine and have
consciousness.)

However we view it, the image of Gaia as a living being is a powerful
one filled with unexplored implications for reimagining ourselves, civilization, Earth, spirit, and our future. What, though, are some of the pitfalls and potentials involved in using this particular image in this way?

In the first place, we must ask ourselves just what we mean when we
say that the Earth is a living organism, or even more fundamentally, what

we mean by the term *organism*. In many ways, our sense of life has become more biological and materialistic than it was for our ancient ancestors who saw physical existence as part of a larger spectrum of being that included non-physical and spiritual aspects as well. For us, an organism is a biological entity that has lost those higher, sacred resonances. We explain its existence and functioning strictly in terms of chemical and physical properties and reactions without reference to spiritual causation or phenomena, which was certainly not the case in pre-industrial cosmologies. Therefore, while the idea of the Earth as a living organism was unquestionably a powerful and pervasive image in earlier cultures, what this meant for our ancestors is not what it might mean for us. We cannot just adopt the idea and assume that we have regained some ancient wisdom. When we strive to image the livingness of the Earth, we do so in a very different cultural context than did those who took for granted a larger view of what it meant to be a living being and who assumed as a given that all life was part of a large sacred dimension.

Can we simply adopt and graft onto our culture our ancestors' notion of a living, sacred Earth? Can we simply accept Gaia as an acceptable reimagination of the Earth? I don't think so, at least not without distortion. We cannot simply dismiss the effects, both good and bad, of two hundred years of modernity nor how they have shaped our cultural consciousness. If our society is moving toward our own equivalent of a more holistic and spirit-filled sense of the world and our role within it (as I believe it is), then our image of Gaia must arise out of our modern and postmodern experiences. We have to think deeply into this idea and live it out in our present context. In effect, we must rediscover the meaning of Gaia for ourselves rather than graft onto our situation an image from history. For we lack the living nature and the mysteries of creation which earlier cultures took for granted and which provided the daily context for their affairs. Without those deeper connections, understandings, and insights, the image of Gaia becomes a novelty rather than a tacit assumption, remaining superficial. It is an idea to think about rather than an idea that creates the context for other ideas. It becomes as an image a suit to try on and wear when it benefits us rather than a body to inhabit and live through.

We are the products of a largely materialistic, rational, technological culture that over two hundred years ago set aside the ancient and medieval perspectives of the Great Chain of Being in which each and every life was sacred and had a purpose, a place, and a meaning. It is this sense of the whole and of the individual as an expression of the whole that we have lost. We have a sense of incarnation but not of co-incarnation—the many ways in which the fabric of our identities are interwoven and interdependent in ways extending far beyond just the human milieu. Thus our definitions of life become reductionist and utilitarian. In business, for example, the emphasis, particularly in recent years, upon the economic bottom line has made us forget that there is a "top line" as well that affirms and responds to the spiritual and holistic value of a person, a plant, an animal, and a place. In such a context, without a natural sense of a larger inclusive wholeness inculcated in our very upbringing, what does it mean to us to speak of the Earth as a living being and to have it mean more than just a scientific curiosity? How do we translate this idea into something meaningful and transformative for our time?

Accepting Gaia simply as a "return of the Goddess" or jumping on the bandwagon of a new planetary animism without thinking through the implications of just what Gaia might mean in our culture can lead to sentimentality rather than spirituality or a rational and imaginative reconstruction of society to incorporate our awareness of a new relationship with its largest and most universal member. By accepting the image of Gaia, particularly by seeing it essentially as a metaphysical or mystical image, we may think we have made a spiritual breakthrough when in fact we are simply indulging a romantic fantasy lacking the power to reorganize for the better our lives and our society. To invoke the "spirit of Gaia" is no substitute for hard-edged, practical political, economic, social, and scientific work to redress the ecological imbalances currently endangering us. If the idea of Gaia is to inform, empower, and sustain our culture to make the hard choices and difficult changes necessary to secure our children's future, then it must be more than just a clever, sweet, or sentimental image. It must do more than hearken back to an animistic past, for even if we were now to determine and accept that the basic premise behind animism—that all things possess spirit and are part of a larger spiritual wholeness and ecology—is correct, we would still need to understand and formulate that new animism in a manner that made sense in our historical context.

I believe Gaia can be an important spiritual idea for our time, but for it to fulfill its potential, we must remember that a spiritual idea is not something we think about but something that inhabits and shapes us. It is like a strand of DNA, organizing and energizing our lives. A spiritual idea is not just another bit of data to be filed away. It is incarnational rather than descriptive, coming alive and revealing its full meaning only when incorporated (made flesh) in our lives through work, practice, skill, passion, and reflection. It becomes part of the architecture of our lives. Being a new icon for worship or ritual is not enough.

However, a deeper question is, do we really need Gaia as a spiritual image? There *is* among many people a strong desire to affirm the sacredness of all life and of the Earth as a whole. However, this desire, it seems to me, is directed not necessarily at coming up with new images of divinity but of restoring value to what has been devalued. It is really a desire to change behavior. What is wanted is to relate to ourselves, to each other, and to the world as a whole as if we all have ultimate value apart from purely utilitarian considerations. If something is sacred, it is assumed to have value beyond its form, its usefulness, its duration, and what it can do or produce. It has value in its being. It is valuable in and for itself.

It is easier to manipulate or exploit something or someone if we have first stripped them of all value except the most utilitarian, changing them from a being into a thing (or, in Buber's terms, from a "Thou" to an "It"). That which is sacred, however, cannot be devalued. So by naming the Earth sacred, we seek to protect it and ourselves from ourselves.

Perhaps, though, this is a form of psychological overkill. Can we not value something just for itself without needing to assign it a special place or condition in the universe? Can we not behave with love and respect for the environment and the biosphere for reasons other than that we consider them to be sacred?

Of course, it can be argued that historically we consider anything that is simply ordinary and not sacred to be fair game for whatever rapacious and manipulative desires we may choose to act upon, which is a fair description of precisely how much of humanity does act, particularly in the modern era. Sacredness could well be considered the protection of last resort, a bulwark against such depredation. However, this attitude does a disservice both to the idea of the ordinary and to the idea of the sacred. It devalues the former and turns the latter into a tool to achieve a particular end. In effect, it makes the idea of the sacred utilitarian, giving it value as long as it can do something useful like changing social behavior.

Using the idea of the sacred in this way, and certainly by using the image of Gaia to give the Earth a special, sacred status, does not address the deeper issues of unwise and uncaring behavior. Solutions to our ecological problems do not lie in renaming the Earth or turning it into a religious icon but in doing the hard inner work of a psychological and social nature to reimagine and transform our everyday assumptions and behaviors. Anyone can be reverent and high-spirited towards whatever they deem to be special and divine, but it is the true task of spirituality to enable us to be honoring and compassionate towards the ordinary, every-day elements of our lives, including each other. Spirituality really has to do with how we behave towards the things of the Earth rather than towards the things of the spirit.

If turning Gaia into a religious or spiritual icon may be inappropriate, it can still be an *inspirational* image. Such an image is like an enzyme. It is not important in itself except as it catalyzes a process. In this instance, the image of Gaia could catalyze our awareness to change and grow in certain ways. In this manner, it could prove to be a very important image for our time.

One effect that this image of Gaia can have is to heighten our awareness of ecological conditions and the responsibilities which we have as members of the environment. It can inspire us to translate theory and concern into practical strategies and actions that can enable our society to live with less destructive impact and greater harmony with the world and lives around us.

A second effect of Gaia is to inspire us to shift the operating paradigm of our society from a purely mechanical one based on classical physics to an ecological one based on biology, thereby putting the phenomenon of life back into center stage in our culture. It inspires us to reform our culture to be truly life-affirming and bio-centered.

A third consequence arising from such a shift of paradigm is to develop a new epistemology, a new way of seeing, learning, and knowing that is not so dependent on reductionist and analytical modes of thought. Gaia can inspire us to develop modes of perception, cognition, and action that are holistic, systemic, symbiotic, connective, and synergetic. We can learn to see the world in terms of patterns and not just separate things, in terms of networks and not just centers and peripheries, and in terms of processes and not just static arrangements. We are inspired to imagine ways of being in the world that are compassionate and co-creative, cooperative and co-incarnational.

Fourth, the image of Gaia does inspire us to explore an eco-spirituality or a way of understanding our own human spirituality within an ecological context. Since many of our spiritual images lead us away from the Earth and towards transcendental states, which can in turn foster a sense of benign neglect or even indifference or hostility towards the physicality of the world on which we live, such a new apprehension of a spirituality that embraces the green and juicy things of nature and the integrity of the land could only help us in fashioning a bio-friendly culture.

An eco-spirituality would have another beneficial effect. Unfortunately, people often use religion as a way of drawing exclusive boundaries between themselves and others, and then using the differences in belief to justify persecution and war. It is common knowledge that one of the psychological manipulations required to enable the average person to fight and kill another in battle is to render the enemy somehow sub- or inhuman; religious conflicts take this to a further extreme by painting the opposition as being outside of God's loving concern and an enemy of the divine. By portraying the Jews as subhuman, for example, the Nazis justified the Holocaust. After all, according to them, real people, real humans, were not being killed, only a strange species of animal that managed to pass for human.

But what about the truly nonhuman? If we can devalue members of our fellow species, how much more easily can we devalue members of other species? By regarding the non-human natural world as being soulless, without spirit, outside the sacred community as defined by a particular religious viewpoint, and therefore outside of any moral consideration, it becomes easy to treat that world as simply a collection of objects whose value lies solely in their utility. But in so doing, we are engaging in another kind of Holocaust. Only this time it is the natural world we have been feeding into the ovens of overconsumption. Only this time, because of the interrelatedness of the biosphere, we are all of us—human and nonhuman—becoming the new Jews, heading toward what philosopher and environmentalist Roger Gottlieb calls "Auschwitz Planet."

A true eco-spirituality, whether termed Gaian or not, that evolved out of the needs and sensibilities of our time, would expand our boundaries of what we consider sacred or, more precisely, the community of value, to include all the nonhuman cosmos (which after all is so very much larger than the human niche within it). By thus enlarging our community of value—the community of what we value and see as kin to us rather than separate from us—we would need to develop a new ethic for how we treat the Earth.

In such an instance, we would also see ourselves in new ways, less as the center of creation and more as a valued and co-equal participant with a contribution to offer that is unique. We would also be inspired to reimagine in a Gaian context the meaning and destiny of humanity. I already mentioned that in his book *Gaia*, Lovelock suggested that humanity might be an evolving nervous system for the planet, the means by which Gaia achieves self-awareness. At a time when our society seems motivated by few purposes higher than endless expansion and economic growth and when humanity seems to have no conception of a purpose beyond itself, this image is striking and refreshing. It would seem to suggest a role that we can play in the world that is more than just the outcome of human desires.

The point is that Gaia symbolizes a way of reimagining ourselves and the Earth in a larger, more holistic and systemic way, requiring us to work out the implications and consequences of such a reimagination.

Gaia *is* an important idea, but I see it as a transitional idea. It is not so much a revelation in itself as it is a precursor to revelation or to new imaginings that can come when that idea is examined and lived more deeply. It must have a chance to settle into our bones. Then who can say just what insights it may open for us and what new patterns of behavior? Its power now is that it can inspire us to be open to and seek out such revelations and changes as a way of becoming a truly planetary species.

Gaia is not the only image we could use to reimagine our relationship to the Earth. Yet, however we perform this reimagination and whatever images we may come up with to guide us into the future, we do so as modern beings moving into a postmodern or transmodern world. The images that will eventually prove most powerful for us in transforming our world will arise from our everyday lives and experiences; they cannot just be imported from our history. They must emerge from our own contemporary acts of embodied imagination, our own act of collective learning that can create a new bond of community between ourselves and the natural order.

⚜

"Our Life as Gaia"

Joanna Macy

Come back with me into a story we all share, a story whose rhythm beats in us still. The story belongs to each of us and to all of us, like the beat of this drum, like the heartbeat of our living universe.

There is science now to construct the story of the journey we have made on this earth, the story that connects us with all beings. There is also great yearning and great need to own that story—to break out of our isolation as persons and as a species and recover through that story our larger identity. The challenge to do that now and burst out of the separate prison cells of our contrivings, is perhaps the most wonderful aspect of our being alive today.

Right now on our planet we need to remember that story—to harvest it and taste it. For we are in a hard time, a fearful time. And it is the knowledge of the bigger story that is going to carry us through. It can give us the courage, it can give us the strength, it can give us the hilarity to dance our people into a world of sanity. Let us remember it together.

With heartbeat of the drum we hear the rhythm that underlies all our days and doings. Throughout our sleeping and rising, through all our working and loving, our heart has been beating steady, steady. That steady, sturdy inner sound has accompanied us all the way. And so it can take us back now, back through our lives, back through our childhood, back through our birth. In our mother's womb there was that same sound, that same beat, as we floated there in the fluid right under her heart.

Let that beat take us back farther still. Let's go back, back far beyond our conception in this body, back to the first splitting and spinning of the stars. As scientists measure now, it is fifteen billion years ago we manifested—in what they call the Big Bang.

There we were, careening out through space and time, creating space and time. Slowly, with the speed of light, in vast curls of flame and darkness, we reached for form. We were then great swirls of clouds of gas and dancing particles—can you imagine you remember? And the particles, as they circled in the dance, desired each other and formed atoms. It is the same desire for form that beats now in this drum and in our hearts.

Ten billion years later, one of the more beautiful swirls of that swirling mass split off from its blazing sun—the sun we feel now on our faces—and became the form we know best. And our lifetime as Gaia began.

Touch our Earth, Touch Gaia

Touch Gaia again by touching your face, that is Gaia too.

Touch Gaia again by touching your sister or brother. That is Gaia too.

In the immediate planet-time or ours, Gaia is becoming aware of herself, she is finding out who she is. How rich she is in the multitudinous and exquisite forms she takes.

Let us imagine that her life—*our* life as our planet—could be condensed into twenty-four hours, beginning at midnight. Until five o'clock the following afternoon all her adventures are geological. All was volcanic flamings and steaming rains washing over the shifting bones of the continents into shifting seas—only at five o'clock comes organic life.

To the heartbeat of life in you and this drum, you too, right now, can shift a bit—shift free from identifying solely with your latest human form. The fire of those early volcanoes, the strength of those tectonic plates, is in us still. And it may well be, if things continue the way they are going, that we will all return for a spell to non-organic life. We'd be radioactive for quite a while, but we are built to endure.

For now and in these very bodies of ours, we carry traces of Gaia's story as organic life. We were aquatic first, as we remember in our mother's womb, growing vestigial gills and fins. The salt from those early seas flows still in our sweat and tears. And the age of the dinosaurs we carry with us, too, in our reptilian brain, situated so conveniently at the end of our spinal column. Complex organic life was learning to protect itself and it is all right there in our neurological system, in the rush of instinct to flee or fight.

And when did we appear as mammals? In those twenty-four hours of Gaia's life, it was at 11:30 P.M.! And when did we become human? One second to midnight.

Now let us take that second to midnight that is our story as humans and reckon that, in turn, as twenty-four hours. Let's look back through the twenty-four hours that we have been human.

Beginning at midnight and until two o'clock in the afternoon, we live in small groups in Africa. Can you imagine you remember? We feel pretty vulnerable; we haven't the speed of the other creatures, or their claws or fangs or natural armor. But we have our remarkable hands, opposable thumbs to help shape tools and weapons. And we have in our throats and frontal lobes the capacity for speech. Grunts and shouts turn into language as we collaborate in strategies and rituals. Those days and nights on the verge of the forests, as we weave baskets and stories around our fires, represent the biggest hunk of our human experience.

Then in small bands we begin branching out. We move across the face of Gaia; we learn to face the cold and hunt the mammoth and name the trees of the northern forests, the flowers and seasons of the tundra. We know it is Gaia by whom we live and we carve her in awe and fear and gratitude, giving her our breasts and hips. When we settle into agriculture, when we begin domesticating animals and fencing off our croplands and deciding that they could be owned as private property, when we build great cities with granaries and temples and observatories to chart the stars, the time is eleven fifty-eight. Two minutes to midnight.

At eleven fifty-nine comes a time of quickening change: we want to chart the stars within as well as those we see in the skies; we want to seek the authority of inner experience. To free the questing mind we set it apart from Gaia. We make conjectures and rules and heroes to help us chart our freedoms to think and act. The great religions of our planet-time arise. At six seconds to midnight comes a man called Buddha and shortly after another called Jesus of Nazareth.

What now shapes our world—our industrial society with its bombs and bulldozers—has taken place in the last few microseconds of the day we have known as humans.

Yet those few microseconds bring us right to the brink of time. And each of us knows that. Each of us, at some level of our awareness, knows that we are doing ourselves in—that Gaia herself, our self, is in danger. And at some level of your consciousness that is why you are here. Oh yes, you may think you are here to heal yourselves on the personal level and find your power in terms of your individual lives. True enough. But we are also here because we know our planet is in danger and all life on it could go—like that! And we fear that this knowledge might drive us insane if we let it in.

Much of the time it is hard to believe that we have come to this—to such an apocalyptic moment. Even those of us who work hard to stop nuclear weapons have trouble really believing that they exist. After the millions of years of life on earth, after the millennia of our civilizations, after Ishtar and Shakespeare and Gandhi and Dorothy Day, we find it hard to credit the fact that we are targeting them at great populations, poising them on hair-trigger alert, leaving them liable to go off on a computer malfunction. . . .

So we are now at a point unlike any other in our story. I suspect that we have, in some way, chosen to be here at this culminating chapter or turning point. We have opted to be alive when the stakes are high, to test everything we have ever learned about interconnectedness, about courage—to test it now when Gaia is ailing and her children are ill. We are alive right now when it could be curtains for conscious life on this beautiful water planet hanging here like a jewel in space. Our foremothers and forefathers faced nothing quite like this, because every generation before us took it for granted that life would continue. Each lived with that tacit assumption. Personal death, wars, plagues were ever encompassed in that larger assurance that life would continue. That assurance is lost now and we are alive at the time of that great loss. It is not the loss of the future. It is the loss of the certainty that there will be a future. It affects everyone, whether they work in the Pentagon or the peace movement. And the toll it takes has barely begun to be measured.

In so-called primitive societies rites of passage are held for adolescents, because it is in adolescence that the fact of personal death or mortality is integrated into the personality. The individual goes through the pre-scribed ordeal of the initiation rite in order to integrate that knowledge, so that he or she can assume the rights and responsibilities of adulthood. That is what we are doing right now on the collective level, in this planet-time. We are confronting and integrating into our awareness our collec-tive mortality as a species. We must do that so that we can wake up and assume the rights and responsibilities of planetary adulthood—so that we can grow up! That is, in a sense, what we are doing here.

When you go out from here, please keep listening to the drumbeat. You will hear it in your heart. And as you hear it, remember that it is the heart-beat of the universe as well, and of Gaia your planet and your larger self.

When you return to your communities to organize, saying no to the machinery of death and yes to life, remember your true identity. Remember your story, our story. Clothe yourself in your true authority. You speak not only as yourself or for yourself. You were not born yester-day. You have been through many dyings and know in your heartbeat and bones the precarious, exquisite balance of life. Out of that knowledge you can speak and act. You will speak and act with the courage and endurance that has been yours through the long, beautiful aeons of your life story as Gaia.

About the Editor

Bill Willers is emeritus professor of biology at the University of Wisconsin-Oshkosh; founder and former director of the Superior Wilderness Action Network (SWAN), a wilderness advocacy organization; editor of *Learning to Listen to the Land*; and author of *Trout Biology: A Natural History of Trout and Salmon.*

About the Contributors

William S. Alverson is a botanist at the Harvard University Herbaria and will soon begin working at the Chicago Field Museum. He is coauthor of *Wild Forests: Conservation Biology and Public Policy.*

Rick Bass is a Montana-based freelance writer and board member of the Montana Wilderness Association. Among his many fiction and non-fiction books is a collection of essays, *The Book of Yaak.*

Bruce Berger is a freelance writer of poetry and prose, and a pianist now dividing his time between Aspen and La Paz, Mexico. His most recent books are *Almost an Island* (1998) and a collection of poems, *Facing the Music* (1995).

Daniel T. Blumstein is a zoologist and animal behaviorist at Macquarie University in Sydney, Australia. He authored the 1995 *An Ecotourist's Guide to Khunjerab National Park.*

Murray Bookchin is a leading social ecologist now living in Vermont. He is author of numerous books and articles on a wide variety of subjects, including dialectical philosophy and revolutionary history.

Charles Bowden is an Arizona-based journalist, author, and contributing editor to *Esquire* magazine. Among his many books is *Juarez: The Laboratory of Our Future.*

David Brower, who lives in California, has been a leading environmentalist for much of this century, having founded a number of organizations, including Friends of the Earth. Widely published, he has lectured all over the world.

John Burroughs was a nineteenth- and twentieth-century naturalist and writer from New York State. His many writings include *Time and Change*. Along with fellow naturalist John Muir he was one of "The Two Johns."

Rachel Carson, who died in 1964, was a marine biologist whose 1962 book, *Silent Spring*, shocked the culture into an awareness of the effects of toxic substances in the environment. Her other works include *The Sea Around Us* (1951) and *The Edge of the Sea* (1956).

Raymond F. Dasmann is a wildlife ecologist at the University of California, Santa Cruz. His books include *Planet in Peril* (1972) and *The Conservation Alternative* (1975).

William O. Douglas, who died in 1980, was an associate Supreme Court Justice from 1939 to 1975 and a champion for the natural world. Among his writings are *My Wilderness: East to Katahdin* (1960) and *My Wilderness: The Pacific West* (1961).

Wayland Drew, who died in 1998, was a teacher and freelance writer of fiction and nonfiction from Ontario, Canada. Among his many books is *Superior, the Haunted Shore*.

David Duffus works in the Department of Geography at the University of Victoria, British Columbia, Canada, as director of the Whale Research Laboratory.

David Ehrenfeld is a biologist at Rutgers University. He is founder and former editor of the journal *Conservation Biology*. Among his books is *The Arrogance of Humanism* (1978, 1981).

Robert Elliot is a philosopher in the Department of Education at Kelvin Grove College of Advanced Education in Brisbane, Australia.

Paul Gruchow, who formerly taught college, and who co-owned a newspaper, is a freelance writer living in Minnesota. His books include *The Necessity of Empty Places*, soon to be reissued.

Guy Hand is a freelance writer and photographer who has traveled and worked in the Scottish Highlands for many years. He is currently working on a book about the forests of Scotland.

J. Donald Hughes is Evans Professor of history at the University of Denver. Among his books is *Ecology in Ancient Civilizations*. He is on the editorial boards of a number of journals, including *Environmental Review* and *Environmental Ethics*.

Lawrence E. Johnson is senior lecturer in philosophy at Flinders University in South Australia. He is the author of *Focusing on Truth* (1992).

Robert Kimber is a freelance writer and columnist for *Country Journal* and *Down East*. He lives in Maine and is the author of *Upcountry* and *A Canoeist's Sketchbook*.

Joseph Wood Krutch, who died in 1970, was a naturalist, journalist, educator, and prolific writer on a variety of subjects. Among his many works are *The Measure of Man* and *The Great Chain of Life*.

Gary Lawless is a poet living on the Maine coast. He teaches at Bates College and is co-owner of a bookstore, Gulf of Maine Books. His latest collection of poems is titled *Caribouddhism*.

John A. Livingston of Ontario, Canada, is professor emeritus of environmental studies at York University and a past president of the Ontario Federation of Ontario Naturalists. Among his books is *One Cosmic Instant*.

Thomas J. Lyon is a former professor at Utah State University and a freelance writer now living in California. He is the editor of *On Nature's Terms: Contemporary Voices*.

Joanna Macy is a California-based author and workshop leader in Buddhist philosophy and the Deep Ecology Movement. Her most recent book is *Coming Back to Life: Practices to Reconnect Our Lives, Our World* (1998).

George Perkins Marsh was a nineteenth-century diplomat and congressman from Vermont. His book *Man and Nature*, a pioneer critique of human impact on the natural world, is a landmark in environmental literature.

Bill McKibben is a freelance writer living in New York's Adirondack Mountains. His 1989 book *The End of Nature* is a major contribution to modern environmental literature. His most recent book is *Hundred Dollar Holiday* (1998).

Ellen Meloy is a freelance writer and artist living in the desert of southern Utah. She is the author of *The Last Cheater's Waltz.*

Carolyn Merchant is professor of environmental history, philosophy, and ethics at the University of California at Berkeley. Among her books is *Earthcare: Women and the Environment.*

John Muir was a nineteenth- and twentieth-century explorer, naturalist, and writer considered by many to be the spiritual father of the environmental movement. His works include *The Mountains of California.*

Arne Naess, professor emeritus of philosophy at the University of Oslo, Norway, coined the term *deep ecology* in 1973 and founded the Deep Ecology Movement. He is currently working on a book of his selected works.

Max Oelschlaeger teaches environmental ethics at the University of North Texas where he is a professor of philosophy and religion studies. He has conducted workshops nationally and internationally. His most recent book is the 1994 *Caring for Creation: An Ecumenical Approach to the Environmental Crisis.*

Sigurd F. Olson, who died in 1982, was a biologist, conservationist, and freelance writer who became a figurehead for wilderness advocacy in the north woods of the Midwest. Among his books are *The Singing Wilderness* (1956) and *Runes of the North* (1963).

David W. Orr is a professor of environmental studies at Oberlin College in Ohio. He is author of the 1994 book *Earth in Mind* and is currently working on a book about ecological design.

Lucy Redfeather is an environmental educator and a Green political activist. She divides her time between Wisconsin and the Rocky Mountain West.

Holmes Rolston III is a theologian and professor of philosophy at Colorado State University. He is associate editor of *Environmental Ethics* and author of *Conserving Natural Value* (1994).

Wolfgang Sachs is a theologian and historian who is project director at the Wuppertal Institute (Germany). He is coauthor of the 1998 book *Greening of the North: A Post-Industrial Blueprint for Ecology and Equity.*

John Seed is founder and director of the Rainforest Information Center in Australia. A writer on issues of deep ecology, he has been conducting Councils of All Beings and similar issues around the world for fifteen years.

Paul Shepard, who died in 1996, was Avery Professor of human ecology at Pitzer College at Claremont College. His many writings continue to influence environmental activists and thinkers. Among his best-known works are *Nature and Madness* and *Myth and Literature.*

David Spangler is a philosopher, educator, and writer living in Washington State. A Lindesfarne Fellow, he is a former codirector of the Findhorn Foundation community in Scotland. His latest books are *Everyday Miracles* and *The Call.*

Henry David Thoreau was a nineteenth-century naturalist and transcendentalist whose journals formed the basis for his book *Walden,* an American classic. He was known for his view that personal conscience supersedes governmental law.

Jack Turner, a former academic philosopher, is a freelance writer and mountaineering guide in the Teton Mountains of Wyoming. He is currently working on a book about that mountain range.

Donald M. Waller is a professor of botany and environmental studies at the University of Wisconsin. He is a past vice president of the Society for the Study of Evolution and coauthor of the 1994 book *Wild Forests: Conservation Biology and Public Policy.*

Howie Wolke is a Montana-based author and wilderness guide. He coauthored *The Big Outside* (1989, 1992), an inventory of remaining wilderness areas of size in the United States, and is president of Big Wild Advocates.

Donald Worster is professor of history at the University of Kansas. His most recent book is *An Unsettled Country* (1994). He is currently writing a biography of John Wesley Powell.

George Wuerthner is a freelance writer, photographer, and ecologist living in Montana. The latest of his many books are *Grand Canyon, A Visitor's Companion* and *California Wilderness Areas: Deserts, Volume 2.*

Acknowledgment of Sources

Willers and Blumstein reprinted from the following issues of *Conservation Biology:* vol. 6, no. 4 (December 1992); vol. 7, no. 2 (June 1993); and vol. 7, no. 3 (September 1993) by permission of Blackwell Science, Inc.

Ehrenfeld reprinted from *Beginning Again: People and Nature in the New Millennium* (Oxford University Press 1993, 1995) with permission of David Ehrenfeld.

Dasmann from "Some Thoughts on Ecological Planning" in Doug Aberley, ed., *Futures by Design.* New Society Publishers (1994).

Redfeather reprinted with permission of the author.

Kimber adapted from an essay in the *Maine Times* (May 14, 1998), reprinted with permission of the author.

Part Two

Drew reprinted from *Ontario Naturalist* (September 1972) with permission of the Federation of Ontario Naturalists (355 Lesmill Road, Don Mills, Ontario M3B 2W8, Canada).

Sachs reprinted from *ifda dossier 68* (November–December 1988) (4 Place du Marche, 1260 Nyon, Switzerland) with permission of the author.

Wolke reprinted from *Wilderness on the Rocks* (1991) by permission of Ned Ludd Books.

Douglas from *A Wilderness Bill of Rights* (1965). Little, Brown and Company.

Meloy from *Raven's Exile: A Season on the Green River* by Ellen Meloy. Copyright © 1994 by Ellen Meloy. Reprinted by permission of Henry Holt and Company, Inc.

Berger reprinted from *The Telling Distance: Conversations with the American Desert* (1990) Breitenbush Books, Inc., with permission of the author.

Turner from *The Abstract Wild* by Jack Turner. Copyright © 1996 by John S. Turner. Reprinted by permission of the University of Arizona Press.

Brower reprinted from *Work in Progress* (1991) Gibbs Smith, publisher, with permission of the author.

McKibben from *The End of Nature* by William McKibben. Copyright © 1989 by William McKibben. Reprinted by permission of Random House, Inc., Bill McKibben, and the Watkins/Loomis Agency.

Hand reprinted from *Northern Lights* vol. XIII, no. 1 (Fall 1997) by permission of the author.

Thoreau from *Walden* 1854.

Bass reprinted with permission of the author.

Livingston reprinted from *The Company of Others: Essays in Celebration of Paul Shepard* (1995) by permission of Kivaki Press, Inc.

Lawless reprinted from *First Site of Land* by Gary Lawless (1990) Blackberry Books, with permission of the author.

Part Three

Hughes from *American Indian Ecology* by Donald J. Hughes (Texas Western Press 1983).

Merchant from *Ecological Revolutions: Nature, Gender, and Science in New England* by Carolyn Merchant. Copyright © 1989 by the University of North Carolina Press. Used by permission of the publisher.

Marsh from *Man and Nature* edited by David Lowenthal. Copyright © 1965 by the President and Fellows of Harvard College. Reprinted by permission of Harvard University Press.

Muir from "Wild Wool" in *Overland Monthly* (April 1875).

Olson reprinted from *The Hidden Forest* (Viking Press 1969) by permission of Craig Blacklock, Minnesota.

Oelschlaeger reprinted from *The Idea of Wilderness: From Prehistory to the Age of Ecology* by Max Oelschlaeger. Copyright © 1991 by Yale University Press. Used by permission of the publisher.

Johnson from "Eco-interests" in *A Morally Deep World: An Essay on Moral Significance and Environmental Ethics* (Cambridge University Press 1991).

Rolston III reprinted from *The Environmental Professional* vol. 13, no. 4, by permission of Blackwell Science, Inc.

Lyon first appeared in *Wild Earth* vol. 2, no. 2 (Summer 1992) (POB 455, Richmond, VT 05477) and is reprinted with permission.

Seed quoted from *Deep Ecology: Living as if Nature Mattered* by Bill Devall and George Sessions (Salt Lake City: Gibbs Smith, 1985), pp. 243–246. Used with permission.

Wuerthner first appeared in *Wild Earth* vol. 8, no.1 (Spring 1998) (POB 455, Richmond, VT 05477) and is reprinted with permission.

Burroughs from *Ways of Nature* (Houghton Mifflin Company, 1905).

Elliot reprinted from "Faking Nature" by Robert Elliot in *Inquiry* vol. 25, pp. 81–93, by permission of Scandinavian University Press, Oslo, Norway, 1982.

Naess from *Ecology, Community and Lifestyle: Outline of an Ecosophy* (Cambridge University Press, 1989). Arne Naess and David Rothenberg.

Gruchow from *Boundary Waters: The Grace of the Wild* by Paul Gruchow (Milkweed Editions, 1997). Copyright © 1997 by Paul Gruchow. Reprinted with permission from Milkweed Editions.

Spangler printed with the author's permission.

Macy in John Seed, Joanna Macy, Pat Fleming, and Arne Naess, eds., *Thinking Like a Mountain: Toward a Council of All Beings* (New Society Publishers, 1988), pp. 59–65.

Index

Abbey, Edward, 103–111 *passim*
Aborigines, 12
  See also Indians
Acid rain, 24
Activists (animal rights), 23
Adam (and Eve), 16
Adams, Bill, 105, 110
Adams, John, 116, 165
Adams, John Quincy, 167
Adirondack Mountains, 24
Aesthetics, 68
Africa, 43, 222
Africans, 16
Agriculture, 29, 38–48, 118, 168–170
Air Force, 34, 36, 37
Ajo, Arizona, 35
Albuquerque, New Mexico, 36
Alligators, 10
Alverson, William S., 22–25
America
  Central, 39
  forests of, 142–143
  North, 12–17, 19–22, 34–35, 39,
    164–170
  See also United States
American Fisheries Society, 63
American Medical Association, 27
Americans, 16–17, 19, 21

Ames, Nathaniel, 166–167
Antelope, 34–38
Anthony, Doug, 196
Anthropocentrism, 23, 56, 90–91, 158,
  188–190
Anthropology, 26
Anthropomorphism (*See*
  Anthropocentrism)
Anti-environmentalism
  (governmental), 90, 96–97
Aphids, 51
Aphis lions, 51
Appalachia, 24
Appalachian Mountains, 94
ARA Leisure Services, 106
Arabia, 30
Archaeology, 38–40
Arches National Park, 112
Arizona, 34–37, 90, 103, 109–110
Arkansas, 18, 146
Arnold, Ron, 62
Asia, 39–40, 59, 62,167
Asians, 16, 167
Astronomy, 26
Audubon Society, 85
Aurochs, 40, 43
Australians, 16
Automobile, 14, 18

Babbit, Bruce, 145

Back of Beyond Expeditions, 104

Bacon, Francis, 168–169

Badger, 150–151

Baram River, 52–56

Barnacles, 50

Barra (Hebridian island), 138

Bartlett, M. B., 168–169

Bartram, John, 10, 15

Bartram, William, 10

Basho, 127

Bass, Rick, 144–155

Bear
    black, 94, 168
    cave, 41
    grizzly, 87, 93–94, 145, 149–152
    habitat of, 87, 151
    in Scotland, 138

Beaver, 138

Beetle (southern pine), 87

Behavior (animal), 45–46

Benton, Thomas Hart, 167

Berger, Bruce, 112–113

Bernard, Claude, 126

Berrill, N. J., 33

Berry, Thomas, 212

Berry, Wendell, 8–9

Bible, 167–168

Big Bang Theory, 221

Bill of Rights, 128

Biocentrism, 22–25, 101, 190

Biodiversity, 56

Biogeochemical cycles, 8

Biophilia, 68

Biosphere, 31

Birds, 33

Birth control, 25

Bison, 11–12, 40, 193–194

Bittern, 138

Blue Ridge Mountains, 10

Blumstein, Daniel T., 58–62

Boar (wild), 138

Bob Marshall Wilderness, 85, 87

Bobcat, 150

Boeing, 9

Bohm, David, 176

Boise Park, 141–142

Boise River, 141–142

Bolivia, 30, 82

Bonneville, Captain Benjamin, 141

Bonneville Power Administration,
    151

Bookchin, Murray, 26–30

Borneo, 53

Boston, Massachusetts, 16

Boswell, James, 138

Bowden, Charles, 34–38

Box elder, 24

Bradley, F. H., 177

Brannon, Ed, 87

Brave New World, 73–75

Brazilian pepper trees, 24

Brazilians, 16

Bridger-Teton National Forest, 85, 87,
    95

Broughton Archipelago, 52

Brower, David, 56, 130–132

Brown, Dee, 119

Bryant, John, 87

Buber, Martin, 217

Budd, Karen, 116

Buddha, 223

Buddhism, 116

Bugs, 11
    See also Insects

Bullard, John, 169

Bureau of Land Management (BLM), 87–102 *passim,* 110, 113
Bureau of Reclamation, 103
Bureaucracy (self-interest of), 99
Burroughs, John, 194
Butterflies, 23

Cabbage palmetto, 10
Cabeza Prieta National Wildlife Refuge, 36–37, 90
Cabinet Mountains Wilderness, 151
California, 35, 142, 151
Callicott, J. Baird, 179–183
Canada, 12–13, 35, 52–55, 149–152
Canadians, 16
Capitalism, 21, 28, 164
Capitol Lake, 112
Capitol Reef, 112
Capra, Fritjof, 191
Carbon dioxide, 14
Caribou, 145, 149
Carolinas (U.S.), 10
Carson, Rachel, 48–51
Carter Administration, 95
Cartesian philosophy (*See* Descartes, René)
Caspian Basin, 39
Cataract Canyon, 111
Cattle
  domestic, 47
  wild, 43, 47
Caucasian race, 167
Celts, 136–137
Centralism, 30
Champion International Corporation, 93, 146–147
Chicago School of Economics, 122
Child's Mountain, Arizona 34

China, 13
Christian Church, 21
Christianity
  and damage to nature, 114–115, 192
  and European civilization, 167
  and Gaia, 214
  and private property, 118–119
  anti-naturalist bias of, 28
Church (Christian), 21
Civilization, 78
Clark, William, 15, 10
Clean Air Act, 119
Clear-cuts, 145, 147
Clearcutting, 94–95
Cleveland, Ohio, 12, 30
Climax conditions (ecological), 65, 94
Coal, 14
Cod, 50
Coleman, Henry, 168
Colorado, 19, 184
Colorado Canyon, 105–106
Colorado Plateau, 104
Colorado River, 35, 103–111 *passim*
Colton, Calvin, 167
Columbia River, 10
Columbus, Christopher, 12
Community and the Politics of Place, 115
Congress
  and frontier expansion, 167
  and immigration law, 13
  and industrial connections, 146
  and public lands legislation, 85–99 *passim,* 119, 150–152
Connecticut, 93
Conservation biologists, 23
Conservation biology, 56, 60
Constitution (U.S.), 114, 165

Consumerism, 28

Consumption (mass), 14

Copernicus, 26

Coronado, Francisco Vasquez de, 11

Corporations
exploitation by, 54, 91–92
and governmental connections, 91, 146
and university connections, 92

Cougar, 94, 149–150

Cove Mallard (Idaho), 142

Coyote, 150

Crane, 138

Creationists, 19

Crowell, John, 88, 96

Culture (human *vs.* wild nature), 180–183

Cynegetics (*See* hunting)

Czechoslovakia, 42

Darwin, Charles, 19, 132

Darwinism, 180

Dasmann, Raymond F., 64–65

Day, Dorothy, 223

Death Valley National Monument, 112

Declaration of Independence, 165

Deep ecology, 191, 208–209

Deer, 11, 22–25, 94–95, 149

Del Webb Corporation, 106, 109

Descartes, René, 164, 176, 179

Desolation Canyon, 106

Detroit, Michigan, 14

Development
industrial, 155–158, 167
sustainable, 52, 55, 65, 179

Devil's Tower, 112

Diamond, Jared, 22–23

Dinosaur National Monument, 107

Diversity maintenance areas, 23

Dogen, 184, 211

Dogs, 43

Domestication, 38–48, 192

Dominy, Floyd, 107

Douglas, William O., 102–103

Dragonflies, 50–51

Drew, Wayland, 73–81

Ducks (wild), 94

Duffus, David, 52–56

Dunn, John, 116

Eagle, 138, 149, 151

Eaton, Reverend H. M., 167

Ecology
and international relations, 178
and revolutionary thought, 26–30
deep, 191, 208–209
shallow, 191

Economics
and language of control, 113–130 *passim*
and management, 9
and resource extraction, 81–82
market, 21
philosophy of, 21–22
versus non-economic values, 68–69

Ecosystems
and moral significance, 179
structure of, 64–65

Ecotourism, 60

Eden (Garden of), 16–22, 50, 164, 168–169

Ehrenfeld, David, 62–64

Einstein, Albert, 17

Elephant, 42

Elk, 15, 94, 138, 149, 192

Elliot, Robert, 194–208
Ellul, Jaques, 75
Emerson, Ralph Waldo, 127
Endangered Species Act, 87, 119
England, 137
Enlightenment, 26, 114–115, 120, 165
Entomology, 49
Environment
  damage to, 19
  engineering of, 195–207
  impact statements for, 91
Environmental Protection Agency
    (EPA), 119
Environmentalism
  and moral order, 19
  as atonement, 16
  shortcomings of, 134
Escalante country, 106
Espy, Mike, 144
Ethics, 69, 212
Europe
  and deforestation, 136–143 *passim*
  and economics, 114
  and philosophy of domination,
    164–167
  as source of industrial goods, 30
  history of, 40
  social organization of, 27
Europeans, 11, 16, 166–167
Evans, Brock, 85
Eve (Adam and), 16
Everglades, 24
Evolution
  organic, 19, 30, 57, 60, 189–190
  Theory of, 26
Expansion (American), 167
Experts, 62–64
Extinction, 19, 189–190

Farmers, 169–170
Federal Lands Policy and
    Management Act, 89, 91
Feedback loops, 8
Finance, 30
Finland, 13
Fir (subalpine), 94
Fire, 12, 24
Fish (and Game) Department
    (Arizona), 37
Fish and Wildlife Service (U.S.), 11,
    36–37, 90, 146–147, 151
Fishing, 40
Flaming Gorge Dam, 104
Flathead National Forest, 87
Florida, 10, 12–13
Fontanelle Forest Reserve, 22–23,
    25
Ford, Henry, 14
Forest of Caledonia (Scotland),
    136–137
Forest Service (Canadian), 53
Forest Service (U.S.)
  and bear habitat, 151–152
  and Champion, Inc., 147
  and Yellowstone, 57
  anti-wilderness policy of, 84–102
    passim
  in Wisconsin, 23
  incompetence of, 121
Forestry, 53, 83–84, 139
Forestry Commission (Scotland),
    139–140
Forestry schools (curricula), 92–93
Forests
  American, 142–143
  history and future of, 135–143
  old-growth, 52–53

Fossil fuels, 14

Foundation for Research on
    Economics and the
    Environment, 120–121

France, 13, 42

French Revolution, 26

Freud, Sigmund, 26

Friends of Glen Canyon, 107

Friends of the Earth, 108

Fritz, Ned, 86

Frost, Robert, 187

Gaelic (language), 136, 138

Gaia, 7

Gaia Hypothesis, 212–224

Galileo, 26

Game and Fish Department
    (Arizona), 37

Gandhi, 223

Gas (natural), 14

General Motors, 9

Genesis, 17

Genetics
    and domestication, 44
    and "supertrees," 93

Germany, 82

Gifford Pinchot National Forest, 84

Glacier National Park, 85

Glacier Peak Wilderness, 85

Glaciers, 39–40

Glen Canyon, 103–111 *passim*

Glen Canyon Dam, 103–111 *passim*

Glen Canyon National Recreation
    Area, 106

Global warming, 24

Goats
    domestic, 40
    wild, 41, 94

God
    and manifest destiny, 166–167
    and subduing the earth, 168–169
    and theory of internal relations,
        177
    and untrammeled earth, 132–133
    and utilitarianism, 192
    See also Christianity

Golding, William, 213

Goldsbury, John, 169

Goodman, Nelson, 127

Goshawk, 94, 138

Gospel (Biblical), 166

Gottlieb, Roger, 219

Government
    corporate connections of, 91
    resource exploitation by, 54

Grand Canyon, 109–110

Grand Teton National Park, 114

Granger, Christopher, 95

Grant Village (Yellowstone National
    Park), 90

Grayback Ridge Roadless Area, 88

Grazing, 57

Great auk, 138

Great Plains, 12

Greed, 20–22

Greek philosophy, 15, 179

Green River, 106

Gros Ventre River, 114

Grouse, 149–150

Gruchow, Paul, 209–211

Gulf of California, 105

Haeckel, Ernst, 27

Hamilton, Alexander, 117

Hand, Guy, 135–143

Hare, 149

Harlan County, Kentucky, 12
Hayduke Lives!, 104
Heaven, 21
Hegel, Georg Wilhelm Friedrich,
   178
Heraclitus, 177
Herbicides, 93
Heron, 151
Hetch Hetchy Valley, 200
Highland Clearances, 138
Highways, 15
Hobbes, Thomas, 164
Holism, 176–178, 218
Holmes, Ezekial, 169
Holocaust, 219
Homeostasis (of Biosphere), 213
Hoosier National Forest, 88
Hopkins, Gerard Manley, 114
Hormonal change, 46
Hornbill, 55
Horse, 41–42, 47
Hubris, 8
Hughes, J. Donald, 163
Humanism, 26, 29
Humans
   impacts by, 59–60
   nature of, 170–173
Hungary, 42
Hunter, James, 137
Hunting
   and gathering, 39
   as control, 25
   by early humans, 40–42
   trophy, 35
Husbandry, 43
Huxley, Aldous, 73–75
Huxley, Sir Julian, 33
Hyde, Lewis, 127

IBM, 9
Ice Age, 40
Idaho, 85, 135, 139, 142, 149, 152
Immigrants, 12, 15
Immigration, 13
Imperialism (by conservationists),
   60–61
India, 13
Indiana, 88
Indians
   Anasazi (sites of), 107
   Cree, 82
   Euro-American domination of,
      166–168
   Nez Percé, 163
   on reservations, 152
   philosophy of, 163
   pre-Columbian populations of, 12
Industrial Revolution, 137, 147
Industrialism, 14, 28, 30
Industrialization, 114
Industrial-political connections,
   95–97
Industry (timber), 53
Insects (control of), 49–51
Instrumentalism, 26, 168–170
Iowa, 24
Iran, 34
Iraq, 13
Ireland, 137
Ishtar, 223
Islands (of wilderness), 64–65
Isle of Skye, 139
Israelites, 15
Italians, 16

Jack pine, 24
Jackson Hole, Wyoming, 114

Jackson Lake, 114
Jackson, Reid, 85, 95
Jackson, Wes, 187
Janzen, D. H., 24
Japan, 15
Jefferson, Thomas
    and America as Eden, 17–19
    and Enlightenment ideals, 115–118
    and extinction theory, 22
    and mechanical instrumentation,
        165
    and the rights of generations, 132
Jellyfish, 50
Jesus, 19–20, 223
Job (Old Testament), 133
Johnson, Dr. Samuel, 138
Johnson, Lawrence E., 179
Johnstone Strait, 52, 55
Jordan, William, 126
Joseph (Nez Percé), 163
Joy, 33–34
Judeo-Christianity, 16, 190

Kalmiopsis Wilderness, 85
Kansas, 19
Kemmis, Daniel, 115
Ken Sleight Expeditions, 106
Kentucky, 12
Keystone species, 23
Khunjerab National Park (Pakistan),
    59
Kimber, Robert, 67–69
Kirtland's warbler, 24
Kite, 138
Kmart, 19
Knutson-Vandenberg (K-V) Act,
    98–99
Kootenai National Monument, 85

Kootenai River, 149
Koreans, 16
Krutch, Joseph Wood, 31–34

Lacewing, 51
La Garita Wilderness, 184
Land Use Regulation Commission, 67
Landfill, 19
Language (economic manipulation
    of), 121–128
Larkin, Philip A., 63
Las Vegas, 12
Law, 22
Lawless, Gary, 159
Leases (oil and gas), 87–88
Leopards (snow), 59
Leopold, Aldo, 32, 91, 101, 103, 191
Lewis, Meriwether, 10, 15
Libby Dam, 151
Limerick, Patricia Nelson, 119
Lion (mountain), 94, 149–150
Little Lake Creek Wilderness, 86
Live oak, 10
Livestock
    as prey, 59
    on public lands, 57
Livingston, John, 155–158
Lobbyists, 95–96
Loblolly pine, 10
Locke, John, 115–116
Lodore Canyon, 106
Logging (to prevent wilderness), 84
Long Island, New York, 18
Lopez, Barry, 151, 187
Louisiana-Pacific Corporation, 85
Love of land, 67
Lovelock, James, 212–214, 219
Luke-Williams Gunnery Range, 36

Lynx, 94, 149
Lyon, Thomas J., 184–187

Mabey, Richard, 139
MacCaig, Norman, 138
Macy, Joanna, 220–224
Madison, James, 115–117, 165
Magnolia, 10
Maine, 68–69, 134, 167–168
Malaysian Government, 54
Mammoth, 41
Management
  and biocentrism, 23–24
  and domination of nature, 164–170
  and ecological simplification, 29–30
  and expertise, 62–64
  and forests, 93, 99–100
  and human presumption, 7–9, 31
  by Stone Age people, 12
Managers (land), 25
Manhatten, New York, 134
Mantis, 50–51
Manufacturers (and nature), 19
Marcuse, Herbert, 75
Margulis, Lynn, 213
Market, 8
Maroon-Snowmass Wilderness Area,
  112
Marsh, George Perkins, 170–173
Marshall, Robert, 101
Marx, Karl, 26
Maslow, Abraham, 26
Massachusetts, 13–14, 168–169
Mayr, Ernst, 175
McDonnell Douglas, 9
McKibben, Bill, 132–134
McTaggart, John, 177–178
Mechanism, 73, 164–170, 176–178

Media (news), 54, 93
Mediterranean Sea, 39–40
Meloy, Ellen, 103–111
Merchant, Carolyn, 164–170
Metcalf, Robert, 49
Mexicans, 16
Mexico, 12–13, 35
Michigan, 19, 24, 63
Midwest (U.S.), 25
Miller, Doris K., 58
Mining, 57
Mining Law of 1872, 91
Minnesota, 19
Mississippi (state), 19, 145–146
Mississippi Valley, 12
Missouri River, 10
Missouri (state), 167
Mites, 11, 51
Modernism, 175–178
Monkey Wrench Gang (The), 104,
  109
Monoculture, 155, 158
Montana, 85, 87, 144–155 *passim*
Montana (school of forestry), 92
Monticello, 10, 165
Moose, 24, 94, 130–131, 138, 149
Moses, 17, 19
Mosley, J. G., 196
Mosquitoes, 50–51
Mother Earth, 82
Mount Saint Helens National
  Monument, 84
Mountains
  Adirondacks, 24
  Appalachian, 94
  Blue Ridge, 10
  Cabinet, 151
  Rocky, 10, 94, 102, 149

Mountains (*continued*)
  Teton, 112
  Zagros, 40
Muir, John, 101, 173, 200
Multiple Use-Sustained Yield Act, 90
Multiple-use philosophy, 57, 91
Mutation, 45
Mystique (of nature), 65
Mythology, 16

Naess, Arne, 208–209
National Environmental Policy Act
    (NEPA), 88, 91
National Forest Management Act
    (NFMA), 91, 98, 100
National forests
  and "displacement habitat" policy,
    152
  and logging roads, 99
  Bridger-Teton, 85, 87, 95
  Flathead, 87
  Gifford Pinchot, 84
  Hoosier, 88
  in Yaak Valley, Montana, 145
  in Yellowstone Ecosystem, 57
  Sawtooth, 85
  Texas, 86
  wilderness preventative logging in,
    84–86
National monuments
  as overmanaged, 112
  Death Valley, 112
  Dinosaur, 107
  Kootenai, 85
  Mount Saint Helens, 84
  sizes of, 117
National Park Service, 89–90, 92, 97,
    106

National parks
  Arches, 112
  as overmanaged, 112
  Glacier, 85
  Grand Teton, 114
  Khunjerab (Pakistan), 59
  sizes of, 117
  Yellowstone, 90, 112, 114, 117, 119,
    192
Native Americans (*See* Indians)
Natural areas, 25
Natural characteristics, 198–207
Natural selection, 44
Natural systems, 59
Nature
  and human culture, 31–33, 180
  and politicians, 19
  as resource, 82
  balance of, 48–51
  humanity's place in, 31–33
  reserves, 57
  rights of, 68
  that was lost, 10–22
Navajo Mountain, 107
Nebraska, 22
Netherlands, 13
New England, 164–170
New Hampshire (school of forestry),
    92
New Jersey, 13, 15, 19
New Mexico, 110
New Resource Economics, 120–121
New World, 16, 20, 22
New York
  city of, 30, 134
  state, 19
New York City Planning Commission,
    27

New York (school of forestry), 92
Newfoundland, 11
Newton, Isaac, 116, 164
Nineteen Eighty-Four, 73, 75, 80
Noosphere, 31
Noranda Mining Company, 151
North America, 10, 12, 15, 17, 19, 35
   See also United States, Canada
North Carolina (school of forestry),
   92
North Dakota, 15
Northeast (U.S.), 25
Norway, 13
Nozick, Robert, 122
Nuclear reactors, 12

O'Toole, Randal, 121–122, 125
Odum, Eugene, 178
Oelschlaeger, Max, 175–178
Off-road vehicles, 64, 87
Ohio, 13
Oil, 14, 30
Oklahoma, 13
Old-growth forests, 52–53
   See also Forests
Old World, 22
Olson, Sigurd F., 173–174
Olympic Peninsula, 142
Omaha, 22
Orcinus orca (*See* Whale, killer)
Oregon, 85
Organicism, 176–178
Origin of Species, 19
Orr, David, 7–10
Orwell, George, 73, 75
Otters, 151
Owls, 55, 145, 149
Ox (wild), 138

Pacific Northwest, 12, 94, 99, 149
Pacific Ocean, 10
Pakistan, 59
Palestine, 39–40
Paperbark, 24
Paramecium, 49
Parasites, 50–51
Parasitism, 47, 51
Passenger pidgeons, 11
Penan (culture), 53–54
Pennsylvania, 13
Perham, Sidney, 169
Permian (geologic era), 51
Pests
   and pesticides, 12, 48–51
   infestation, 29
Peterson, Maxwell, 88, 96
Philosophy, 26, 55
Phoenix, Arizona, 35–36
Pigs, 55
Pinchot, Gifford, 101
Pine (ponderosa), 142
Pittsburgh, Pennsylvania, 30
Planning, ecological, 64–65
Plato, 137
Plymouth Colony, Massachusetts,
   143
Polistes, 51
Political-industrial connections,
   95–97
Political Thought of John Locke (The),
   116
Politicians (and nature), 19
Pollan, Michael, 127
Pollution, 19
Population (human), 13, 15
Posner, Richard, 122
Powell (Lake), 103–111 *passim*

Powell, John Wesley, 111
Power, Thomas Michael, 127
Prairie dogs, 11–12
Predators, 50, 57
Prey, 57
Prigogine, Ilya, 175–178
Private property (as social value),
    117–119
Pronghorn (*See* Antelope)
Protozoans, 49
Public lands, 57, 58, 94, 114–115
Purchaser Credit Roads, 99
Puritanism, 168

Quaker (Pennsylvania), 10
Quotas (timber and road mileage), 97

Railroads (and land acquisition), 117
Rainbow Bridge, 105, 107–108
Raine, Kathleen, 191
RARE II, 95
Rat, 45
Rathbun, Jim, 85
Rationalism, 26
Ravens, 151
Reagan Administration, 95
Reagan, Ronald, 86
Real estate (U.S. history), 117–118
Reason, 74–79
Recreation, 53
Red Sea, 39
Redfeather, Lucy, 65–66
Reductionism, 176–178
Redwood trees, 142
Reformation, 175
Regeneration (of trees), 23
Regionalism, 30
Regulation, 22

Reindeer, 41–42, 138
Renaissance, 26, 175
Reserves (nature), 57
Resources, 81–83
Resourcism, 33, 64, 69
Restoration (environmental), 65,
    195–207
Rhinoceros, 41
Rights of nature, 68
Riparian areas, 94
Rivers
    Boise, 141–142
    Colorado, 35, 103–111 *passim*
    Columbia, 10
    Green, 106
    Gros Ventre, 114
    Kootenai, 149
    Missouri, 10
River Rouge Plant, 14
Roads (logging), 94, 96, 99–101, 145,
    153
Robson Bight, 52–53
Rocky Mountains, 10, 94, 102, 149
Roe, Frank Gilbert, 11
Rolston, Holmes III, 179–183
Romans, 15
Roosevelt, Theodore, 101
Roshi, Robert Aitken, 191
Roszak, Theodore, 191
Rothschild, Michael, 121
Russia, 42

Sachs, Wolfgang, 81–83
Salmon, 114
Salt Lake City, 111
Salt River Range Roadless Area, 88
Sand County Almanac (A), 91, 103
Sarawak, 54

Sawtooth National Forest, 85

Scale insect, 51

Science
  and irreverence, 126
  and management, 9
  and mechanism, 165–170
  and sustainability, 55–56
  and values, 193–194

Scotland, 135–143

Seashores, 30

Seed, 188–191

Seidman, Mike, 23–25

Selection (genetic), 44

Seral (stages and species), 94

Seton, Ernest Thompson, 11

Shakespeare, William, 223

Sheep
  domestic, 40
  wild, 41

Shepard, Paul, 38–48

Sierra Club, 19, 85, 106, 145

Sirmon, Jeff, 96

Sleight, Ken, 103–111 *passim*

Sloths, 17–18

Smith, Adam, 21, 115, 165

Smith, Seldom Seen (*See* Sleight, Ken)

Smith, Vernon L., 121

Snyder, Gary, 127, 185

Society of American Foresters, 92

Songbirds, 23

Sonora (Mexico), 35

South (U.S.), 99

South Dakota, 10

Southern Utah Wilderness Alliance, 89

Spaceship Earth, 7, 9

Spangler, David, 211–220

Spirit Mound, 10

Spirituality, 211–220

Statue of Liberty, 15

Steel production, 30

Stegner, Wallace, 69

Stone Age (people of), 12

Students (views of), 64–65

Sturgeon, 145, 149

Subdivision (of land), 30

Subsidies (road), 99

Succession (ecological), 65, 84, 94

Supreme Court, 108

Sustainability, 63, 155–156

Sustainable development, 52, 55, 65, 179

Sustainable growth, 7

Sweden, 13

Sweet clover, 24

Sweetgum, 10

Swiss, 16

Synergies, 8

Systems (biological), 63

Tao of Physics (The), 191

Tao Te Ching, 191

Technocracy, 75–78

Technology, 8–9, 14, 20, 114

Teton Mountains, 112

Texas, 86, 109

Texas Committee on Natural Resources, 86

Texas National Forest, 86

Texas Wilderness Bill, 87

Thailand, 13

Theology, 212

Thompson, William Irwin, 214

Thoreau, Henry David, 31, 144, 185–186

Tibetans, 16

Tigris-Euphrates Watershed, 40
Timber (harvest of), 58
Tin, 30
Totalitarianism, 29
Tourism, 53
Trout, 145–151
True, Dr. D. N., 169–170
Tucson, Arizona, 35–36
Turkey (country), 39, 42
Turner, Jack, 113–130

United Kingdom, 13
United States
  and urbanization, 27, 30
  in history, 12, 15
  relationship to nature, 22, 60
  See also America
University-corporate connections, 92
University of Montana Foundation, 93
Upland Island Wilderness, 86
Upper Colorado River Project, 108
Ural Valley, 151
Utah, 103–111 passim, 185
Utah (school of forestry), 92
Utilitarianism, 168

Values, social, 115, 193–194
Vancouver Island, 52
Van der Post, Laurens, 75
Victorian Era, 26
Virginia, 13

Walden, 144
Waller, Donald M., 22–25
Wasatch Mountain Club, 108
Washakie Wilderness, 87
Washington (state), 84–85
Washington, D.C., 58, 145

Wasps, 52
Waterfowl, 11
Watt, James, 87
We, 73–74
Wendy's (restaurant), 19
West (intermountain), 94
Western Culture, 175
Whale, killer, 52
White, E. B., 134
White, Richard, 119
Whitehead, Alfred North, 175–191
Wilderness
  advocacy of, 56
  and governmental bureaucracy,
      83–103 passim
  and moral order, 19
  as human need, 173–175, 184–187
  as islands, 64–65
  as legal entity, 113
  as matrix, 64–65
  Bob Marshall, 85, 87
  Cabinet Mountains, 151
  characteristics of, 200–206
  destruction of, 73–80, 117–118, 146,
      169
  Glacier Peak, 85
  in Yellowstone, 58
  Kalmiopsis, 85
  La Garita, 184
  legislation for, 114
  Little Lake Creek, 86
  management of, 100, 192–194
  Maroon-Snowmass, 112
  species dependency on, 93–94
  travel in, 209–211
  Upland Island, 86
  value of, 132, 179–183
Wilderness Act, 86, 91, 185

Wilderness Preservation System, 102
Wilderness Society, 19
Wildlife ecology, 84
Wildness, 38–40, 69, 132, 184–187
Wilkinson, Charles, F., 127
Willers, Bill, 56–62
Williamsburg, Virginia, 16
Williamson, Oliver, 122
Wilson, E. O., 68
Wisconsin, 24
Wise Use Movement, 62, 115–116
Wolke, Howie, 83–102
Wolverine, 145, 150
Wolves, 55, 114, 119, 138, 145,
    149–151, 168, 193–194
Worster, Donald, 10–22, 119

Wuerthner, George, 192–194
Wyoming, 19, 85, 87, 114, 118,
    130–132

Yaak Valley, 145–153 *passim*
Yellowstone Ecosystem, 56–58, 60–61
Yellowstone Falls, 112
Yellowstone National Park, 90, 112,
    114, 117, 119, 192
Youngstown, Ohio, 30
Yuma, Arizona, 35

Zabriskie Point, 112
Zagros Mountains, 40
Zamiatin, Eugene, 73–75
Zimmerman, Michael, 176